guido loero

boundary conditions and global management

1975

Distributed by
ACADEMIC PRESS NEW YORK AND LONDON
A Subsidiary of Harcourt Brace Jovanovich Publishers

Distributed by
Academic Press, Inc.
111 Fifth Avenue, New York, New York 10003
Academic Press, Inc. (London) Ltd
24-28 Oval Road, London N. W. 1

Printed in Italy

To Ca' Foscari
University of Venice

CONTENTS

FOREWORD

Our research is the fruition of practical experience in managerial concern. Therefore, our work may be said to be inspired by reflection, undertaken with professional commitment, based on a direct observation of facts. We entered the working world while the Italian economy was being organized according to a corporative system and remained conditioned by necessarily autarchic objectives.

We thus gained true insight, even if during emergency circumstances, into what complete intervention across the board, by an authoritative state, really means.

Once the war was over and tragically lost, we witnessed the reconstruction that mobilized all available forces through a sum total of efforts which - while often disorganic - were nevertheless capable of leading to an effective production recovery.

This took place within a private enterprise regime, affected only minimally and sporadically by public powers. In the following years, development became more orderly and coherent and eventually gave birth to what came to be known as the economic miracle.

Its real matrix, as in many other countries, was the economic aid supplied by the United States; the resulting achievements followed with astonishing swiftness. Our wonder arose, perhaps, out of our having underrated the importance of the work force in relation to production means.

What one thought of as an economic miracle, may more correctly be defined as an extraordinarily fast economic recovery achieved through the availability of means that allowed full exploitation of an already present labour potential. We were thus able to reach quite a respectable standard of living, especially with regard to private consumption, in a nation like ours so harshly disrupted by the world conflict.

Undoubtedly, many of the essential public services and indispensable infrastructures were ignored.

Similarly, social reforms that ought to have mirrored the deep changes in our political system were either overlooked, or failed to satisfy the needs for which they were created, especially if we compare them with their strictly economic counterparts. At present, Italy and so many other nations in the world are consequently undergoing a period of crisis.

Public opinion believes that the responsibility lies with the political and economic executive classes. The accusations, as usual, manifest themselves in the most diverse ways. Where someone may affirm that "each country has the government it deserves", others reply " it is not difficult to govern the Italian people... it is downright impossible".

We believe ourselves to be a nation that cannot bring itself to realize that our collective interests are pre-eminent and fundamental to the attainment of personal desires. We may simply mention the widespread indifference to public property and its administration, and the fact that people always see its inadequacies and never attempt to formulate constructively committed judge-

ments based on a real desire to participate as members of the collectivity. This diffidence, this mistrust that alienates the people from their government, finds its earliest and precise origin in a historical context. It appears to us that the Italian crisis, not unlike those experienced by many other countries, has much deeper roots. It finds its concrete expression in the contradictions that arise from a capitalist kind of socio-economic reality which is pressurized by agitations derived from Marxist ideology.

This dualism arises out of a profound dissatisfaction with a living pattern that tends to satisfy always decreasingly essential needs at the expense of environmental conditions and man-nature relationships.

A monopolistic capitalism cannot abandon its consumerist policies, since on them rest its most intimate "raison d'être".

All new entreaties, and public pressures for a kind of life that is not exclusively tied to economic values, have stabilized at a wishful-thinking level and do not constitute, at least for the time being, a programme capable of providing a valid alternative. After all, a widespread and clear awareness of these desires is still far from being shared by other than a small minority.

The majority, while feeling the pinch and showing some intolerance, channel their anger into a desire to increase their material living stardards. This, in turn, leads to an exasperated and exasperating consumerism which, as we already mentioned, fosters the growth of a monopolistic capitalism. The paradox contained in such a situation, where the constantly growing material requirements are coupled with an ever-increasing desire to return to a simpler existence inside a more congenial environment, is mirrored by the reality of everyday life. It is especially conspicuous in the world of work; daily frustrations that cause intolerance and lack of incentives are no longer a prerogative of the working classes they are forcefully affecting wider and more qualified categories.

Our work, necessarily, reflects the impact of this reality that characterizes the times we live in.

This explains why our thoughts on man, the environment, and enterprise are motivated by a deep desire to verify the possibilities available to reach a compromise, however gradual and temporary, that will allow us to reconcile the elements of the paradox.

After all, no system has ever been able to condition the individuals within it for an indefinite period of time.

All systems or societies will always contain opposing forces capable of modifying, little by little, the conditions and structures of power. Such forces combine with external influences at a certain, inevitable stage, and provoke an evolution toward more valid and contemporary forms within their general context.

We therefore envisage in the environmental changes surrounding man today an extreme call for new and effective solutions. These solutions, from our point of view, will not ripen in a violent manner, that is, they will not foster or cause a mood for revolution in the traditional sense of the word.

The two great social classes considered to be the protagonists of revolutions throughout history have been the capitalist bourgeoisie, that Marx held responsible for the '89 revolution, and the proletariat.

Yet today we feel that the capitalist bourgeoisie in its struggle to preserve the established order, attempts to restrict rather than expand its economic activity seeking when necessary the indirect support of other social forces. The present industrial proletariat in the Western world, is a class whose aims and aspirations are clearly reformist in character and hence not revolutionary in the least.

Revolutionary fervour subsists among those minorities who for various motives, including psychological and cultural ones, feel themselves out on a limb and ostracised from the rest of society; they are consequently forced to act in accordance with these feelings and attempt to blend their programmes with those that appear to be the interests of more numerous and important social classes.

These minorities, however, cannot entertain the possibility of realizing any concrete success. We therefore arrive at the conclusion that the revolution, as we have understood it through history, will not take place in Europe or at least in the Western world.

We consequently believe that the socio-political reality will gradually transform itself and the present economic crisis will accelerate its evolution.

CHAPTER ONE

PRICES, COSTS AND SYSTEMS

THE PROBLEM'S BOUNDARY CONDITIONS

The developments that lie ahead for us in the future are determined by the economic and political events that shape our present.

A situation that in the western world had always been considered conducive to an increase in our well-being is now revealing its more ominous consequences, and the swiftness with which these consequences react with one another leads to a new situation that we may do well to consider with the utmost urgency. The changes involved do not concern solely production methods or the quality of products, but also the very concept of social co-existence.

The whole environment is being transformed, especially if we take into account with this keyword both natural phenomena and the cultural context which is, in itself, simultaneously the cause and effect of our material condition.

A gross oversimplification of the problem would rest on one notion — man's survival — or at least that of civilization as we have understood it until now and as we understand it today.

We are obviously referring to the development crisis as it has been so effectively defined by the recent research conducted on the subject at M.I.T. The crisis implies a new and drastically different relationship between economics and the environment from what it was until a few decades ago.

The context of economics is a reality that emerges out of the night of time, with the earliest collective efforts of social Man; as a discipline, it finds its birth with the appearance of currency, it evolves until the seventeenth century as a non-dogmatic science (the state no longer being sole arbiter of civil existence) and therefore as an independent field of thought.

Today, economics cannot be other than an all-embracing discipline that concerns itself with the whole world in a contemporary perspective and in the light of an ecological parameter.

Economic laws, like biological ones, are rigid and as such determine the interlocking relationship between economics and ecology - a relationship of very recent acquisition, yet a very ancient reality.

A wide field of research has thus come into being, matching the complexity of reality within which man operates.

Barry Commoner invites the economist and the environmentalist (in the broadest sense of the word) to join together and reach, if necessary, beyond their respective spheres of action.

Clearly, we must consider this double aspect of the environmental and economic reality as the objective of an interdisciplinary research, since what we are concerned with is not the solution of transient trends but rather a series of basic, long term choices.

While looking for solutions to cope with a future that is now almost indistinguishable from the present, we must realize that social reality is still determined by political ideologies that find their origin in the philosophies of classical authors. We thus recognize the great relevance assumed by the thoughts formulated by Ricardo, Malthus, Smith and Marx. This social reality will be analysed in a manner apt to demonstrate how these authors of the classical era, have fulfilled and are fulfilling a determining role in the political ideologies of today and tomorrow. We have verified the consistency of our exposition with the yardstick afforded by a Marshallian approach.

The allocation of resources inside a market system is the pivot on which neo-classical thinking revolves. Obviously, when resources are scarse, a study on the causes may be profitably conducted on Marshallian lines, since its concerns are directly involved with the workings of the market system as supreme determinant of these phenomena. The beliefs held by this genius of economics are deeply rooted in microeconomic matters and this is another reason in his favour for someone who wishes to experiment hypotheses as to what the future holds in store for enterprise and other economic organisms. Besides, Marshall himself states that "Economics is not a body of truths; it is merely an instrument to be used in the quest for those truths".

An explanatory survey, conducted on the interaction between supply and demand, aiming toward the formulation of prices, leads one to notice that the structure of marginal costs influences the supply curve, and thus the total costs will increase with the growth in the number of the additional units manufactured in a given product. Clearly, in a system with limited resources, the costs will increase as research into their true nature is carried out by extending and amplifying the concept of cost in itself.

Another fundamental observation, due to Marshall, is that long and short periods are not absolute temporal concepts, but rather that they are relative to a given circumstance and to a particular phenomenon. Actually, this observation is nothing but a redefinition within the sphere of an economic concept of time in which Marshall envelops all industrial undertakings. From our point of view, we consider Marshall's economics as spatial and not multiform. First of all because he has always given priority to concrete facts and to the verification of theories by means of actual related events, and secondly because, like him, if we probe the market and the interdependence of the various operators, we cannot do other than place enterprise in a system that involves the whole world with respect to necessities, contacts and exchanges of all kinds.

Reaffirming that on the structural plane our method is indebted to Marshallian teachings, in as much as we have employed certain mathematical configurations solely for the purpose of using a more concise form of expression, as thorough confirmation of the conceptual pattern with which we have always begun. In fact, Marshall warns us of the pitfalls in the use of mathematics as a pure instrument for research, since its findings can only be validated through concrete proof. Naturally, everyone receives from one's own existential context the information required for an explanation of reality. It follows that Marshall should also be tied to his times and hence should not realize sufficiently the dangers of oligopolies, and should in turn believe that State intervention in economic affairs ought to be avoided. However, he does realize that governments can influence the allocation of resources through taxation and other means. This is what we have pointed out in an attempt to accord to economics the life-giving drive of private initiative, a self-regulating market and the efficiency and automatisms of pivotal choices.

All this appears to be in perfect line with the question that Marshall poses himself on whether the satisfaction of collective needs may not be better arrived at through a more careful allocation of resources and production methods.

The American theoretician, in opposition to the classics, underlines as a fourth production factor the role of capacity for enterprise. Throughout our work, the importance given to this capacity for free enterprise is postulated as the "conditio sine qua non" of the validity of any system. We cannot, however, overlook the fact that in the world of work there are conflicts and agitations that find their truest expression in business concerns, where they also demonstrate, most clearly, their influential power and determining vitality.

Evidently, one of the most important changes will be that concerning the future character of enterprise. We feel we have emphasized it sufficiently. The direction followed by enterprise will be affected by inevitable transformations, it will find its level in the means available, in the ends it decides to pursue and in the methods it will, finally, employ. Even if dimensionality, technologies, instruments and apparatus were to remain identical, there will always have to be a marginal differential element.

As they are perceived by the public at large, the evaluations that interfere with the initial impetus and the consequent realization of facts in the industrial life and with the delineation of an image of industry are subjective ones. This type of psychological consideration generates a complex of ultimately decisive factors, whether we look upon them as internal phenomena or as external reality affecting business concerns from without. They can therefore present themselves as intrinsic parts of any aspect, determining the individuality of enterprise in a very, accurate, historical context. By saying this, we wish to emphasize the importance of the environment as a psychological constant.

Man lives in the present, with memories of the past and hopes in the future, while enterprise lives projected into the future by means and acquisitions achieved in the past.

The present tense is meaningless for enterprise especially when one is required to draw up a final balance sheet.

The social and economic value of manufacturing concerns lies, in fact, in their future prospects, and it is in this context that we must consider the value of their inception and measure their success in relation to their capacity to affect the external psychological reality.

In the light afforded by our present reality, in the breathtaking evolution of technology and traditional customs, the multifaced environmental factors must be considered as a vital tool in the understanding of the nature of enterprise as it is inevitably being transformed in conjunction with the dynamics of social reality.

Never as much as today has each value, each choice needed to be re-appraised in full concordance with the hustling and hurtling transformations that are taking place in the world. With regard to the methods and systems available for the comfort of ultimate decisions, we have come down on the side of a more pragmatic approach that does not submit to an absolute faith in abstract and sophisticated procedures.

In our research, we have examined both industrial and more general costs, against a socio-economic background, and we have treated environmental factors as conditioning variables, in order to carry out an examination of the functions fulfilled by costs and prices, within the sphere of a research geared to establish the perspective of economic phenomena in their ever-changing context. The interpretation of this context, and the prospects for coherent programming, have given rise to a market hypothesis. Through a schematic comparison between free market, programming systems and mixed systems, we have reached a hypothesis which we have turned in to a working proposition that may suitably be termed "transition system". To illustrate it, we have adopted mathematic expressions as well as conceptual paradigms for a hypothetical reality of the future and we have also made use of the series of examples achieved by the Dynamic System.

We have also tested a number of critical theories against the possibility of a prediction or the capacity for intervention, inasmuch as the efficacy of any system is never beyond a reasonable doubt; this aspect of reality, so stimulating in itself, must not be ignored.

Considering man as the fundamental factor in all the problems we shall examine, we have taken into account certain considerations that are not strictly classifiable as economic, because we regard man as the protagonist in all structures and in any programme.

Under this profile, the differences between the western world and that of largely Marxist orientation are not as relevant as they may appear at first sight.

The pre-capitalist era was characterized by a long period of production stagnation caused by the unchanged quality of traditional working methods that enhanced the development and improvement of working means.

Both peasants and craftsmen transmitted to their successors unaltered instruments, techniques and habits they had acquired in their working lifetimes.

The essential clause for manufacturing goods-that is, the independence of production means, like capital for instance, from the worker-was thus lacking.

As Marx accurately suggests, when we have-on the one hand-paid work, and-on the other-capital, the production means will rapidly improve themselves.

Naturally, this takes place at the worker's expense, that is to say, of his professional skill.

Increase, in production at first, has certainly done nothing to improve the condition of a human majority on the contrary, development has initially been paid for with the greatest sacrifices in respect of working conditions, previously held in a precapitalistic society.

Yet, today, as we have already mentioned, on a practical plane and from a purely materialist point of view, the goals, in both the so-called western and socialist worlds, are identical, with similar prefigurations on the "profile" of the dangers toward which an indiscriminate development is leading us.

Together with these affirmations, we need to completely demistify the old distinction between European and non-European societies, since it is the former's feeling of superiority over the latter which has been taken as the fundamental historical justification for colonialism.

Only by recognizing the validity inherent in each culture and the right of each to differ — to be free of oppression in the guise of "improvement" — shall we be able to prevent the mistakes made by the European countries (and most industrialized nations today) from recurring in third-world countries.

Management, as a whole, implies the necessity to take into consideration and evaluate the quality of life, which demands a careful study of the problems pertaining to a new economic and scientific consciousness.

Our present knowledge and the amplitude of the subject are such that it is virtually impossible to assimilate all facets of this knowledge satisfactorily.

However a considerable effort is being made, in both cultural and scientific worlds, to rediscover the unity of vision which would permit us to apprehend the problems within a global context.

Given the environmental situation toward which we are heading and the global context in which man finds himself at work it is vital for each management to be conducted according to the most diverse principles. The relevant problems that must be examined will be approached from a more cognisant and socially responsible standpoint.

THE TECHNOLOGICAL CONSTANT

The integral nature of our present form of civilization contrast strikingly with that of its more fragmentary antecedents, which had themselves arisen out of clusters of more limited, independent factions. It involves the whole world, it contains all races and creeds; there are no states, nations or geographical regions that are excluded from its general context.

Technology has reached the stage of being taken for granted; and third-world countries aim to avail themselves of the structures already operational in more "civilized" nations.

What presents itself to us is a mass civilization in which the conflicts among various ideologies are perhaps, in the long run, destined to dwindle away. In this situation, it is hoped that the problems are studied as part and parcel of an all-embracing system, rather than as sectarian questions, bearing no relation to one another. The interaction between various elements of such complex problems are in fact becoming gradually more apparent.

In socio-economic matter, too, all programmes must be considered as details on a wide canvas that portrays a most complicated interplay of motifs. At present, all sciences tend to examine systems rather than single phenomena.

It remains in fact clear that any aspect of a given situation cannot be understood unless it is analysed and perceived as an intrinsic part of a wider reality which conditions and is, in turn, conditioned by it.

When confronted by a very complex system, it is worth one's while to establish a starting point from the given data and subsequently to set out certain goals for oneself. Reason and logic must determine or influence the location of starting points and final targets.

The attitude with which the researcher approaches the problems determines whether we shall have an over-all pragmatic or an over-all rationalistic attitude.

The elaboration of conceptual paradigms, be they mathematical or of any another kind, may be considered part of a rational method in which all phenomena are conceived as being representable and hence as objects worth studying.

A more pragmatic approach, on the other hand, has for a long time determined almost all our operative choices. But, by now, reality has become so complex that there is no harm in trying to resolve the great problems that face humanity, through research procedures that sometimes may appear a little sophisticated, but that, ultimately, alone can generate the premises for sound choices.

The concept of measure is common in our human experience. We express ourselves in terms of measures, whether we talk of how heavy or how tall a body of matter may be, or how long it takes to cover a given distance. A determined size is compared with another which then stands for one unit.

In geometry, measuring units can be arbitrarily fixed only in relation to a linear class; for all other classes one adopts units that depend on linear unity. In physics and in other scientific fields, certain units may be conventional, but more complex measuring processes make use of sets of units that find meaning only in relation to conventional integers. Research aiming at a form of measure, relative to man's needs, which we shall term "vital economic space", implies the knowledge of a function that expresses an unknown size in terms of other physical shapes that may be measured. The kind of research we thus propose differentiates itself from that conducted at M.I.T., which is geared toward a global evaluation of the present world situation in relation to the well-publicized development crisis.

It is necessary to bear in mind that, while the different elements are constantly recycled, the energy expended cannot be used again. We only need to remind ourselves that consumption of matter implies consumption of energy and thus, in order to save energy, we must save matter.

We recall Einstein's formula $E = M \times C^2$

Naturally, man, like any other living creature, needs a certain amount of energy for his daily sustenance and activity. The interdependence between metabolism and energy can only lead us to conclude that if we want to find a measure which will define the bare necessities for man, we may only find it within the energy required.

We have arrived at the concept of an energy budget, which is certainly not an original concept, since it has often been employed in relation to the different countries to measure the entity of energy expended in terms of tons of coal or other fuels. A kind of measure that indicates the bare necessities of each single individual must be referred to several parameters; it could be indicated by the necessities each man has for a period of one year to avail himself of a determined quantity of water, space, minerals, oxygen, food, all this for a kind of life that was familiar before the industrial revolution.

Logically, a man's needs for a year will be expressed in terms of variably measured quantities, thus relative to the measure adopted for that type of consumption or that kind of matter: we shall, then, arrive at a series of objective quantities. It follows that the sum total of those figures will not be the same for men of different geographical affiliations. But, if we will refer ourselves to the fundamental material principles, such as those governing clorophillian vegetables, it follows that it will be possible to arrive at a sufficient degree of abstraction which will enable us to generalize according to the measure mentioned.

In order to carry out this argument satisfactorily, or at least to control the possibility of the prospects we have outlined, we would require a deep knowledge of biological, chemical, botanical and zoological laws. However, allow me to conclude by reasserting that, if we gained an approximate measure of a man's needs for a year of life, we should be able to calculate the suitability of a determined ecosystem given the entity of the human population.

The reduction of receptivity, implicit in the more evolved forms of life could, then, be calculated on the basis of greater consumption and relative pollution. This would provide us with certain data, even if vague, on the various possibilities of an ecosystem according to the standard of living desired thus a correlation between standard of living and population-size would be made available.

In actual terms, the present population on the globe is partly crowded — a minority — into a relatively small and prosperous area, and partly scattered — a majority — over a large area, suffering an insufficiency of energy resources necessary for life. If we take into account the number of people in the world, we must agree that it is too high to survive on existing natural resources, and the underdeveloped countries, where we have a very high density of population with compartively little natural resources, suffer the most.

ECONOMICS AND THE ENVIRONMENT

In order to formulate a pattern relative to a configuration of industrial costs pertaining to our context, we must, first of all, briefly outline the difference between market and planned economics and underline what is divergent between the two systems.

We have brought to bear on the following analysis De Maria's work, other relevant writings by Bruno Finetti, and a critique on Keynes compiled by Paul Mattrick with the aid of several acute observations introduced by William Barber.

Let us begin by examining the implications arising out of a collectivist policy; its general aims may be said to be the annihilation of private capitalism and the firm adherence to scientific pragmatism.

In the history of the Soviet Union, we have witnessed the destruction of bourgeois banking and credit organizations and an experimental form of bureaucratic economics; the so-called NEP (New Economic Policy). These specific aims have been advanced by means of the two Five-year economic programmes of 1928-1932 and 1933-1937, and before that with the adoption of NEP. A certain freedom was hence given to economic activity, in the hope of the consequent formation of a kind of gigantic collectivity.

In order to reach these goals it had proved necessary to: mobilize all natural resources of the country; create a new complex of ambitious industrial schemes and great enterprises for the distribution of agricultural produce; train the masses to a kind of work unknown in Russia to that day; import goods, machine tools, and raw materials from foreign countries. While this process was under way, it was necessary to keep the standard of living at a low and uniform level throughout the nation. The results achieved in Russia, due to uncompromising effort and to certain favourable starting conditions, have been superior to those obtained through the industrial revolution in England and Germany. The methods employed in such a transformation are therefore of primary importance.

The internal set-up in the Soviet Union has been planned in such a way as to involve each undertaking with a trust composed of similar enterprises. Each trust controls production programmes, proper utilization of raw materials, personnel departments, the quantities produced, all buying and selling; they are managed by a commercial organization made up of various unions.

The industry pays for the production factors it buys, and earns money for the goods it sells; if the manufacture of products requires only semi-skilled labour, the respective firm is credited with an account at the relevant branch of a State bank. Cash payments are required (thus no banking transfers) only to deal with wages, small expenditure and agricultural products. There are two vital moments in the politics of production: the first fixes the quantity of what is to be produced, the second allots to each single factory a certain number of production factors.

In a free economy, raw materials, work and capital are obtained by entrepreneurs and distributed among the various factories according to the financial market (the supply and demand laws) so that each contractor aims to minimize his expenditures and maximize his gains by resolving a whole host of problems. In a planned economy the central authorities establish what must be produced in accordance with political considerations, and therefore plan a set of priorities according to predetermined objectives.

The accomplishment of these aims, for instance the creation of a great industry as is the case in Russia, has required enormous sacrifices with regard to the standard of living.

Within the basic plan, each industry must act in economically sound terms, and in this respect, there are no criss-crossing motives since both have for their goal industrial efficiency. The only point of divergence is constituted by the establishment of priorities and of the quantities to be achieved of each type of product, in relation to the programmer's choices.

In the free market, prices are determined by the interplay between supply and demand, and these, in turn, determine the quantity of products; hence the *price has a truly productive function*. This is not applicable to a collectivist economy, where prices are only a useful indication for accounting activities; in this case they prove relevant to production and distribution (of goods) techniques, but they do not determine them.

Prices are not the result of a bargaining process; they are decided by programmers to give an incentive or a disincentive to the consumption of a certain production.

Therefore, a prices policy in a planned economy must be tied to a principle of perfect adherence to the programme's requirements. Prices are always higher than costs and the profit margin may be fixed at a high or at a low rate according to whether it is desirable to boost or to let it stagnate for a certain period.

The prices fixed by programmers determine the standard of living of the population. If the production of a particular good increases, following a greater efficiency found in a free market system, it will — or at least it will tend to — increase the standard of living.

With all likelihood, in a programmed system, with the sums of production factors saved in a particular sector will be transferred onto another, raising the price of a good or leaving it where it stands, even if the production costs have decreased. Profit, in a collectivist system, cannot be understood in other terms than those of a regulated saving according to the direction indicated by the budget's plans. The budget technique requires an enormous articulation of jurisdictions, and responsibilities, and hence, of decisions.

We have examined, schematically, the differences between market and planning economy. We must now analyse the actuality of a mixed economy as it presents itself in our own world.

Programming in France is undertaken seriously and scientifically. Holland and the Scandinavian countries have adopted a programming policy. In the United States, the Tennessee Valley Authority has organised and restored to health a state that had been subjected to great hardships. Finally, Great Britain and Belgium are both leaning toward the adoption of a coherent programming. Consequently, the E.E.C. seems poised in readiness to direct European economics toward a harmonious development while awaiting a collective, unifying, and wholly European programme.

All this has come into common practice, overcoming prejudices and conservative thinking that had underlain a widespread reticence to even classify as planning what, in reality, was a precise and accurate form of programming inspired by Soviet experiences.

The tendency to discourage certain approaches, theories, and studies related to an economic programming, reduces programming itself to a predictive function and obscures the determining factor of production and therefore of the economic reality.

A summary of R. Frisch's thought would read something like this: whoever wishes to explore the problem of an economic programming must above all free himself of the tendency to think in monetary terms. The world of facts will survive whatever economic institutions man has and will create. Beyond dividends, interests, money, etc., it is necessary to consider concepts like daily bread, clothes, health services and the most effective methods to carry out what has been proposed. We must formulate our plans with the deep awareness of these most real and immediate considerations.

An inevitable consequence of a programmed system, even in free market fostering nations, is the widening of public intervention in the private sector. In practice, where gigantic, industrial dimensions prove necessary, a free market can only be conducive to the formation of oligopolies. These organisms may, solely, act under the incentive of profits, disregarding any social consideration, even in the most vital areas. It is necessary, in this eventuality, to remove the characteristic discretion of private enterprise in more delicate matters and rely on state intervention.

This comparison among free, mixed, and programmed markets cannot be made from a morally and intellectually neutral vantage point, even if many believe in the necessity of a judgement that reaches beyond these concepts and that releases all resolutions from all non-technical exigencies. Nevertheless, we cannot ignore the presence and value of an interaction between traditional customs and the history and general thoughts at the basis of economics. Thus, the concept of profit as the ultimate end expresses a fundamental principle of liberalism which cannot be accepted any more within a social and political context; hence, an absolute code of values cannot exist in the economic reality. To hypothesize a market made up of economic subjects, to be solely directed by the quest for profits, has ceased to be a scientific hypothesis because it no longer reflects life's praxis.

We must examine the so-called mixed economy, because it

is the kind most practiced in the western world. The analysis of the situation determined in the socio-economic field by such a praxis is indispensable for an accurate evaluation of possible alternatives.

Keynes' theories have exerted a great influence both in America and Europe. Let us remind ourselves of certain focal points in his theories and on the consequences which have come to light since, thanks to more recent studies on the subject.

The State acts upon demand by purchasing certain goods from private industry, and finances its expenditures with taxation. Thus, the State changes the character of production and transfers to the public sector the money it has acquired in the private sector. The State may also avail itself of credits and run its balance into deficit.

Now, if we acknoweledge that by capital is meant production methods and raw materials (physical capital) and money (liquid capital) we must arrive at the conclusion that during State production the profits and the redemption of public debts are covered by money borrowed from the State itself. Yet, since the money coming from taxation, like that borrowed from public funds, is private property, the State has to pay a certain interest rate. The production induced by the State increases, but so does the public debt.

We have seen that the State, when it purchases and encourages production, does so with private money; the repayment can only occur through taxation or through new credits. Consequently, the State's commitments can only be refunded by taking money away from private revenue. In other words, the private sector of the economy sees its capacity for capital gains curtailed. The tendency in the last few decades has shown an increase of production induced by governments over that activated by private enterprise.

Let us now assume that the present "favourable" situation in the western world is superficial. Ultimately, the assertion of a mixed economy demonstrates that capitalism would already be in a critical position if there had been less State intervention.

We may do well remember that Marx, before anyone else, cast some doubt on the durability of economic laws by proposing to link them to well-defined situations or historical moments. It is because of this that the particular situation which will soon come into existence (i.e. the worsening of present conditions) will generate a whole new set of economic concepts.

State intervention will involve an increasing degree of controls, and it leads to a conniving relationship with the great body of private capital. Oligopolies will have to be subjected to State intervention (which is in principle ill-received by them), on the condition that — for their self preservation — the great industries' requirements be spared from government handling.

As we have already had cause to mention, State intervention is ever increasing in degree and range, and with it the need to take away increasing portions of wealth from the private world. This process leads to an increase in global production, but the costs of this production will never be capable of counterbalancing retail prices, because, naturally, public works or similar forms of expenditure cannot be sold to anyone.

Investments, be they private or public, incentivate production and consequently increase the global revenue; the same cannot be said for capital which may only be increased by an increase in profits. Thus, State intervention will never manage to increase global capital, and the following fall in the interest rate of capital envisaged by Marx. Subsequently, we may affirm that Keynes' theories do not provide a solution in that they merely postpone the crisis Marx had already predicted.

Since a mixed economy is ultimately only an expedient measure it would be necessary for the State to reach beyond all contradictions and seize control of the whole economy. In other words, since we are watching the sunset of private enterprise, it seems pointless to attempt to keep it alive through artificial means.

Mixed economy, and therefore State intervention, may be regarded, in some respects, as a move towards socialism; it must not be simply a source of profit for a handful of powerful capitalists.

Social provisions, like old age and invalid pensions, and social services in general are not examples of redistributed wealth, nor can they be said to be social provisions at all since they do not increase the worker's income.

The most obvious proof for the above is given by the fact that workers' contributions are far greater than the expenditure carried out in their favour by the government.

An economy only partly programmed by the State remains, in actual fact, a capitalist economy impoverished by a State intervention that relies increasingly on private revenue. Thus all tenets of a capitalist economy are preserved.

Wages are production costs, each wage increase not directly related to greater productivity will diminish the income capacity of the capital and in the long run will impair private enterprise in its struggle to renew its equipment and keep abreast with the times; it follows that the State will, consequently, be able to rely less on private revenue.

A mixed economy, in which great private capital must ultimately play an important role, must establish and maintain a high standard of living for its working classes. Yet this means a decrease in profit margins and an increase in non-profitable production. If the quantity of unremunerated work, provided by the working classes and accumulated by the capitalist ranks, increases with sufficient speed to enable a transformation into capital with only a certain amount of paid over-time, salaries will grow and, assuming all other circumstances to be constant, unremunerated work will decrease proportionally. But, at the point at which the overtime that feeds the capital is cut, either partially or completely, we have a reaction: a minor part of income is capitalized, accumulation is paralysed and the upward movement of salaries is halted.

Revolutionary rumblings may still exist in the poorer strata of society, that is excluded by all factions, because they could precipitate certain situations militating against the will and interests of all, even the so called left-wing parties.

The well-being of the proletariat has made it possible for us to think that the authentic proletariat with its influencial power has been reduced. To clear our minds, we must take into account the fact that in a capitalist society there are only two basic classes of people: on the one hand a minority — a ruling class — capable of taking or influencing all important decisions, and on the other hand a majority — that of the workers — having no opportunity to make decisions.

Public intervention increases global income and production, but in the long run it will weaken the private sector on which a mixed economy relies. A weak condition within the capitalist system may therefore be generated by a fall in the interest rate of capital and may lead to a wearing down of a possible proliferation of capital through that part of unused profits.

The environmental considerations already mentioned cannot but accelerate a movement towards a planned economy, the procrastination of which can only cause further damage.

Hopefully, air, water and earth pollution — the physical afflictions which will alarmingly continue to manifest themselves due to insane management — will galvanize and spur on the will to act and put right a deteriorated concept of the State.

We need to move outside the schemes of profit-based society, in order to appreciate the concept of a planned existence. This may involve great sacrifices, because the price we must pay is a reduction of the quantitative level of life, which will, naturally, not be accepted warmly or serenely, especially until such measures reveal an improvement of qualitative values.

Since the considerations that precede this situation are nothing more than an attempt to insert certain prospects which will come about with the worsening of the situation inside a socio-economic context, we must point to the fact that Marx, who predicted with extreme clarity certain very relevant aspects of the present reality, did not realize the bearing which an indiscriminate growth of population would have on our existence. Malthus' thought, not only because of the great insights he demonstrates on the matter, acquires today a significant position in the resolution of these problems.

RISING PRICES AND PROGAMMING

Let us examine a product P in its distribution phase, and let us account for price behaviour as it is affected by an increase of its wholesale and distribution costs, which for clarity's sake will contain two groups: work, services and other costs. Let us then divide the variables that act towards a formulation of retail price in a levelling off situation, into three categories: work, services and other costs, cost of materials. In order to preserve the balance, any increase in any of these factors will be reflected by an increase in retail prices.

If the greatest increase refers to materials, let us say 30%, a surcharge of 30% is not sufficient to restore the balance; the surcharge must be greater.

Let us illustrate this with an example: the margin on the final price, that will absorb labouring and servicing costs, will be 30%: the corresponding surcharge on wholesale price will be 42.85%; if the material's price is subjected to an increase of 30%, the labouring costs of 10% and the services of 20%, we shall have the following graphic equation, in relation to the previous data.

Labour	20	10%	22
Services	10	20%	12
Goods	70	30%	91
Price	100		125

The charge of 34 units for labouring and services would be sustained by a 27.2% ($\frac{34}{125}$) final margin on a business round of 125.

But in a general system plagued by rising prices and diminishing purchasing power, an increase of 25% in the volume of business is unthinkable. Therefore, if the volume of business remains 100 units, the configuration of equilibrium will be: labour 22; services 12; material 66; which involved a gross, final margin on retail price equal to 34% and a surcharge margin equal to 51,51%.

Hence, if the volume of business remains stationary, an increase of 8.66% on materials and one of 42.85% on surcharges will have to be added on the original figure in order to reach a 30% distribution costs, the purchasing power of the monetary unit has gross margin. This means that for consumption, because of distribution costs, the purchasing power of the monetary unit has lost an extra 4% in the passage from wholesale to retail prices. This 4% is not a shopkeeper's whim; it expresses the necessity for a 4% increase on the final price which is, in fact, the condition dictated by the required equilibrium.

If, due to the loss of purchasing power, a restriction of the volume of business is deemed necessary, the equilibrium clause will have to be obtained trough a greater proportional increase.

Luxury goods are therefore destined to suffer greater proportional increases, while in the sector where consumption, due to the economic trends, will widen, smaller or negligible increases may be installed. For instance, if fewer cars and more bicycles are sold, the latter's price may even decrease in price as a reflection of greater demand. This is an example of the market's self-regulating characteristic. Any external intervention on the government's part will create certain displacements and cause either sizable losses or unjustifiable profits on the logic functioning of the market. If intervention is necessary, it must be handled in such a way as to intervene in one fell-swoop with a planning system that is alien to profit laws.

By applying a programming that implies certain precise choices in the allocation of production factors (at the rightly-called « production level », including distribution and services) the uncomfortable conditions potentially to be suffered by the operative classes in such a situation will be avoided.

It is true that the self-regulating capacity of the market could lead to the annihilation of activities where costs prove unsustainable, but then we must ask ourselves if in the present situation it is possible to let the market take care of itself with its unquestionable efficiency, but also with the social aggravations it may cause in the cold, inexorable light of logic which controls its dynamics.

The market's self-regulation resolves itself with the closure of activities that are becoming obsolete, but behind the word « activity » there is a whole host of large and small industries, of productive sectors unable to operate that are incapable of catering for their employees. Public intervention, unemployment benefits, would affect the remaining industries that are somehow able to survive, to the point of causing further disruption.

Nevertheless, it is certain that the cost of goods and services from private or public sectors would increase, causing another chain reaction of positive feedback loops in the sphere of a decreasing net profit pro capita.

In the modern society, the displacement of production factors from one sector to the other is neither smooth nor fast and not only because of the aforementioned problems.

A situation of rising costs proves to be the only logical hypothesis, in the present context of an indiscriminate development and its consequences in the rarefaction of resources, pollution, and the ever rising population. All these factors contribute to the formation of a new reality which, as we have already had cause to mention, has been determined by a worsening of the ecological balance.

Having ascertained that costs will never follow any but an upward spiral, we shall attempt to provide a mathematical pattern and describe the relevance of its processes through the dynamics of the systems. This will point to the availability of certain short-term choices in the production of goods and services on which one could channel certain production factors which would then be diverted from less important goods and services. On the practical plane, we must take into account the basic necessities of an individual, or rather, for the agglomerations in which most of the population in this particular historical phase are living: that is, family, neighborhood, city. The basic necessities — food, clothes, health and public services — must be given priority. Once these fundamental necessities are satisfied, we shall be able to establish the importance of a second category of need.

Logically, when one talks of choices, one does not limit the field of reference to goods or services that may be currently found today, but to those most economically sound goods and services. In this sense, it is clear that, if — for instance — there is a sufficient quantity of labour and a insufficient quantity of physical energy (electricity, oil, coal) it would be absurd not to put to good use human energy in order to save rapidly extinguishing raw materials. Inevitably, the tendency, in various sectors, is to develop labour—intensive techniques more than those which would require a great expenditure of energy. Whether a certain programming is adopted or not, the extinction of reserves continues and the production of energy implies consumption of matter and consequent pollution. Thus, logical steps must be taken.

It is important to remember that pollution, great natural disasters and the food shortages have not caused widespread panic outside the affected areas. Apprehension, however is rapidly spreading because, whenever something occurs outside the natural laws, it sparks off a chain reaction resulting in great discomfort and speculative anxiety throughout the world.

The industrial areas that have experienced gas discharges, for instance, have created terror among surrounding populations. The famines that have developed in various African states are still inspiring mass migrations. Even in Italy, floods have occurred which, although nothing has been done to prevent their recurrence, have created massive outcries and apprenhension. Paralyses of services, like those resulting from blackouts of electric energy and gigantic traffic jams, have repeatedly taken place even in the United States. Since the dynamics of these phenomena follows an expotential development, it appears absolutely clear that cause for alarm will grow in intensiy and frequency.

A WORKING MODEL FOR INTEGRATED COSTS

The analysis of a given product's cost and of a certain service has, so far, been compiled within an environmental context where resources would be practically infinite; the absorbing capacity of the biosphere, with regard to all forms of pollution, would be unlimited. We have also considered the increase in population and that of a global industrial production as subjects in a linear development rate while the manufacture of goods and pollution occur according to an exponential development.

Under the guiding light of these theories, the costs of a product have been examined without criteria faithful to the actual situation. Considering the rapid exhaustion of raw material reserves, the availability of wich is limited to a few decades (or less), these goods must have a value that cannot escape a more precise application of the supply and demand laws.

As a matter of fact, thinking in terms of the interests relative to one or two generations, and not referring to certain value schemes that may be adopted and carried out by our progeny, an inventory of the availability of raw material must lead us to a completely different set of values that stems from extraction costs and transcends quantitative evaluations, but which finds its primary justification in a comparison between availability and quantities required to cope with the present situation.

Over and above all this, we must take into account the unquestionable value that certain natural elements employed in a productive process possess over others which have, until now, been disregarded cost-wise.

For instance, oxygen plays an active part in any combustion process. In a relatively small quantity it is part of the common goods but can be used in any quantity, even dangerous for man's health and for the preservation of the *habitat,* without the need for an enterprise that avails itself of its services, to pay relevant benefits to the sorrounding community.

Water, consumed in large quantities by many industries, is paid at a rate that is even inferior to that charged for household consumption. The present price for water is laughable, when we consider the actual knowledge on the scarcity of resources still available.

But, still in the combustion field, the carbon dioxide, sulphuric acid, sulphur monoxide and carbon monoxide produced or expelled by industry cause serious damage to people and material that are not in any way protected by the industries concerned.

Now, it is a well known fact that where seemingly far more trivial events are concerned (for instance the opening of a window) the wheel of compensation is often set spinning.

In pratical terms enterprises avoid calculating the cost relative to the use of natural materials available to everyone, and pay a pittance for other resources. Moreover, the damage produced by polluting by-products, when considered at all, is always under-estimated.

Even if we admit that the concept of compensation is in direct conflict with current moral attitudes, and that we may hope for the emergence of alternative technologies, the fact remains that the cost of the inevitable conversion of obsolescent equipment is not taken into consideration and this because of the realization of the danger of the situation that is developing.

Thus, we shall witness a paradoxical situation which — while it clings to classic economic patterns — will adopt the conflicting theories of modern industrial logic.

The installations, perfectly in time with profit laws, will have to be substituted for reasons that have nothing to do with industrial costs as we have understood them until now. The transfer towards more expensive but less polluting technologies will be inevitable. Naturally, products will continue in their actual price spiral and their consumption will therefore be influenced by the gross income per capita that will be registered in unpredictable situations.

We may do well to give relevance to the open contradiction that exists between the two instances. On the one hand the desire for a cleaner environment, on the other, the aspiration towards more and more goods at always lower prices. In the same way, the desire for a decrease in working hours and the application of technologies which, in order to create less pollution, will have to rest on a labour-intensive rather than a capital-intensive principle. What must be closely observed together with the reticence to change which will be encountered in industrial mentality, is the same reticence on the part of the consumer.

One of the most important determining factors is the objective of efficiency in the quest for determined projects and their realization, be they inspired by private or public concerns. When all is said and done, we may define a certain type of economics, a certain way of looking at management in both public and private factories, as theft economics; and emergency economics may be a suitable term for the undeferrable necessity to evaluate all conditions in the light of a more careful and thorough costs analysis. We shall therefore term « conventional costs » those calculated with the guidance of emerging situations. Let us adopt the following symbols:

P	=	industrial product
P1	=	a first kind of industrial product
P2	=	a second kind of industrial product
Pn	=	a nth. kind of industrial product
GP	=	global product
C	=	conventional cost of P
C1	=	conventional cost of P1
C2	=	convenitonal cost of P2
Cn	=	conventional cost of Pn
mc	=	raw materials costs
cc	=	capital cost
lc	=	labour cost

Thus $C = mc + cc + lc$

IC	=	integrated cost of P
IC1	=	integrated cost of P1
IC2	=	integrated cost of P2
ICn	=	integrated cost of Pn
IS	=	integrated cost of S

IS1 = integrated cost of S1
IS2 = integrated cost of S2
ISn = integrated cost of Sn
Rmc = coefficient to be attributed to mc in order to evaluate the incidence on the raw material's prices in its foreseeable exhaustion. Logically Rmc is directly proportional to the eventual increase of industrial employment of said material;
Rcc = increase coefficient in capital cost. Said coefficient will be directly proportional to the difficulties related to the accumulation of the capital itself for the maximalization of savings employment in consumer goods instead of investments goods. It will increase with the inevitable decline of the installation's efficiency;
Rlc = increase coefficient of labouring cost easily ascertainable given the trend of the recent period;
Rpp = coefficient to be attributed to C in order to evaluate the incidence on the cost of P by the pollution damage translateble in a cost for the collectivity that is rightly attributed to the product itself.

Thus IC = (mc x Rmc + cc x Rcc + lc x Rlc) x Rpp.

The difference between (mcn x Rmcn + ccn x Rccn + lcn x Rlcn) x Rpp and C1n will have to be expressed, on a practical level, under the guise of tax employed by the authority as a discouraging or stimulating factor.

The range of disincentives may be enormous; in such eventuality it will make alternative technologies or the employment of other raw materials, more economically viable. In a non-capitalistic society, it will form an evaluative criterion for the various technological choices.

Technologies and raw materials that are considered as conventional cost would surely be chosen according to a profit logic or according to that of efficiency, depending upon whether we are confronted with a capitalistic or socialist backdrop. Clearly, the cost calculated with the coefficient IC has the sole aim of constructing a filter which would be employed only if pressures from either public opinion or political considerations will cause a reshuffling of priorities.

We now wish to introduce, within the examination of prices, a new parameter which expresses the utility of a product in the context of an already termed emergency situation.

To this parameter we shall give the symbol FC and we shall call it the factor of convenience.

This factor of convenience FC will hold the same significance in a same consumption category, for example: food consumption (FC factor constant); heating (FC factor constant); transports (FC factor constant for any type of transport).

$$\text{The ratio } \frac{IC}{FC} = \text{differentiated cost}$$

By differentiated cost we mean that cost wich re-establishes the balance with integrated cost in such a way as to create the possibility for operative choices on a selection level of products and services.

The complete formula is:

$$\frac{IC}{FC} = \frac{(mc - Rmc + cc \times Rcc + lc \times Rlc) \times Rpp}{FC}$$

and we notice that white the numerator has for each single kind of product variable values, the denominator FC is equal in all cases. In this sense the fact FC will be introduced to evaluate better the cost among categories of products or services (Plate 1 and). FC will instead be a variable in the field of cost of a product belonging to a same category.

Evidently, we are confronted with two kinds of research if FC will be chosen for all products of one category; $\frac{IC}{FC}$ will determine the size of the quantity of production factors destined to that category, assuming that we find ourselves in a situation based on limited production factors.

If FC will be different inside one category, $\frac{IC}{FC}$ by indicating a different cost for each good or service, will influence the consumption of that good or service. In the former instance, FC constant for all the goods belonging to one category, the cost $\frac{IC}{FC}$ will be especially useful to programmers; in the second case FC as variable for each good in one category will determine the size of the consumption for each good.

Clearly, the programmer's choice, determining in the final analysis the quantity of the production factors for each consumption category, will influence the cost of the good because Rmc, Rlc, Rcc, Rpp will be affected by the programmer's choices.

We could, therefore, term FC' the constant factor of convenience chosen for a planning of goods and services with discrimination on a categorical level: in such a way certain priorities among categories come into being; FC'' the variable convenience factor, chosen to determine the costs of single goods, in the sphere of action of one specific category: consequently a priority principle would also be established with regard to each single good.

Be that as it may, the scale of values among which we may choose the parameter FC could revolve around the unit 1:

..0,01.. 0,02.. 0,03.. 0,99... 1... 1,01.. 1,02... 2..: 3

We wish to point out that the factor of convenience has nothing to do with the concepts of marginal, total utility or ophellimity etc. Such debated concepts are foreign to the parameter FC exactly because it is a parameter.

The studies on utility conducted by Smith, Pareto and Gossen will be contained in the overall evaluation, in the programmer's field; the decisive conclusions of which will inevitably be influnced by socio-economic and hence political presuppositions that constitute the inevitable starting point for the human decisions which concern collective problems.

We must understand this formula as a means for alternative choices, taken in accordance with a criterion by the finalities that society has proposed and thus foreign to the single preferences and the traditional market indications.

The formula, in fact, demonstrates that we must take into account the reality of raw materials, used in each single product, the quantity of work required, the costs of production equipment, not forgetting the decree of damage that the product may generate, all this in relation to the product's utility in the context of a determined situation. Society, therefore, substitutes herself for the individual in those matters where the general indifference of the individual allows a free choice on the part of the individual himself.

If we must choose among many containers, we would choose the one which requires easily available raw material which in turn produces the least pollution once discharged, needs a minimal amount of manual labour, requires minimal investment in equipment, and that would possess a high degree of convenience. Plastic containers, for example, would have to be disregarded because of their high polluting properties (in this case FC = zero); so would elaborate or carefully aesthetic containers because of their excess work required; or containers too fancy in form and colour since they would need particularly costly equipment and finally very small containers since their convenience is too low with respect to a medium size container.

We may even have to ban production of mineral water, because packing, labour and distribution costs, together with the inevitable pollution, would be far too high when compared with those of a simple aqueduct managed and guaranted publically; especially because the convenience in the great majority of cases is very low per bottle in relation to domestic delivery, in particular because of the fact that most bottled water is sold in places already serviced by the aqueduct thus the convenience of the bottle would be higher only in few and far between instances, which would any case probably be better served by varying the size of reservoirs.

The aforementioned working formula involves computations and estimates; now, while computations way be right or wrong they are nevertheless conducted on the basis of a formally exact logic, whereas estimates are always determined by choices. Thus the formula summarises and places in evidence the changes in a process that in the final analysis is that of a development which today is indiscriminate, but which tomorrow will, by necessity, have to be set against the logic of a system and a programming committed to the evaluations of a small group that will have to take decisions on behalf of the collectivity. Such decisions will be base in innumerable data, elaborated by prodigious computers, but ultimately interpreted according to a convenience-yardstick that will vary in connection with the goals set out by society, the priority scale chosen in the sphere of consumption categories. This presupposes the formation of a new cultural pattern, of new conceptions, and of the increased disjunction between decisional power and delegating mass, where an effective forma of delegation may still exist.

Computer programming will have to be managed with effort parallel to that inherent in production or distribution processes. In order to carry out all types of studies and research, this science — born of a transmission problem with Hertzian waves — will have to improve and refine itself much more. On it rests the foundation for all future programming possibilities.

The discussion on choice-availing parameters and on the prospects facing this future society is already taking place in countries and environments and among men who are not only spatially but ideologically distant from one another.

Man is made to achieve happiness. It is not necessary to own a car to be happy. In the transport field as well as in other areas, we shall look for rational solutions. Beyond a certain material limit, material things cease to have great meaning. Hence life concentrates itself in its cultural and moral aspects. We want to lead a complete, multilateral, rich and happy existence, one in which man may be able to express all his real values. This is what gives meaning to life and value to a people.

GRAPHS PERTAINING TO A COSTS ANALYSIS

RELATIVE ASPECTS OF MARKET AND PROGRAMMED ECONOMIES WITH RESPECT TO THE SWIFT EVOLUTION OF THE ENVIRONMENTAL SITUATION

PATTERN OF MANAGERIAL COSTS ACCORDING TO THE SYSTEM'S DYNAMICS

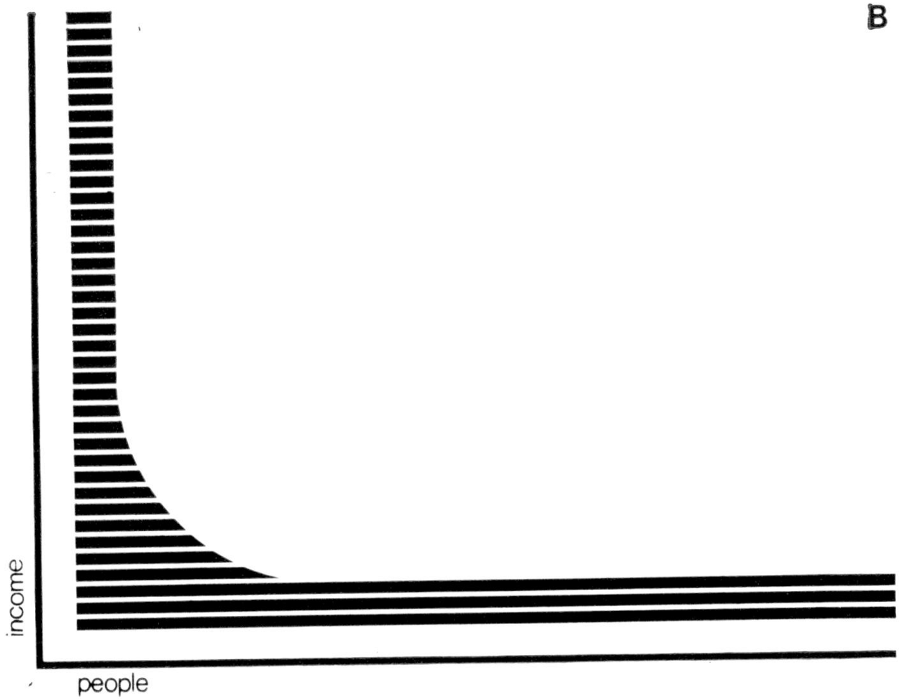

Graph A: Income pattern per person in a feudal type of society, where a very small number of individuals enjoy a very high income and a multitude a pitifully low one.

Graph B: A similar type of society, where the multitude earns a slightly higher income and the élite enjoys an even higher income than that of the previous situation.

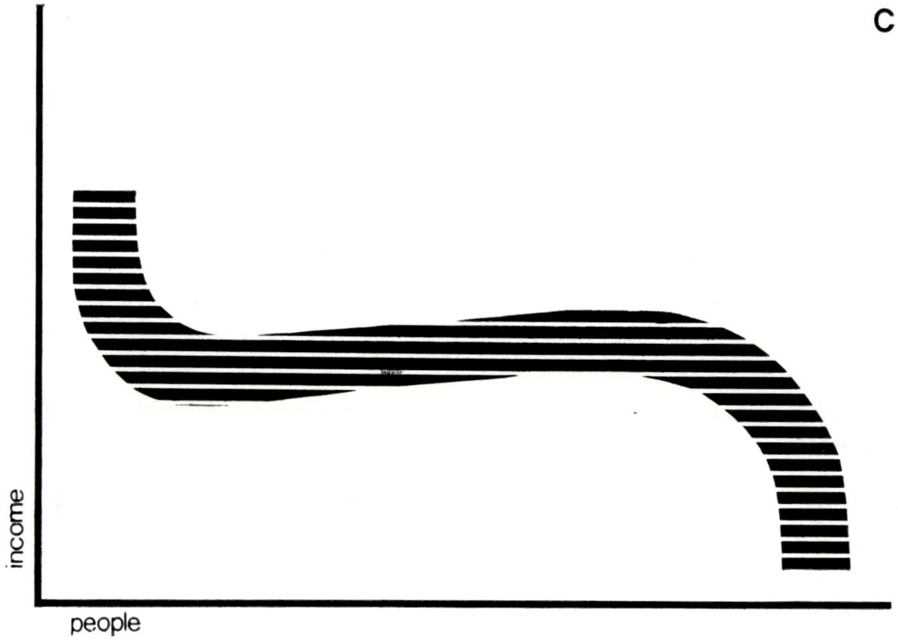

Graph C: The majority enjoys a fairly good income; a considerable number of people an excellent income; some a high income, but inferior to those in the B situation; a small number of individuals find themselves in considerably different conditions.

Graph D: Establishes a purely hypothetical situation where income is equal for everyone.

Let us term NIpc the net income per capita of a single individual and np a number of individuals. Let us choose a high NIpc with respect to a certain society and hence referable to a few people;

$$\frac{\text{NIpc}}{\text{ICn}}$$

we shall have a value which may be either inferior or superior to one unit; clearly, if the ratio is inferior to the unit, the product is practically out of everyone's reach; if it is superior to 1, and we shall refer to this number with X, $\frac{1}{X}$ will provide us with the amount of NIpc absorbed by the purchasing of that product.

An examination conducted on np groups classified on a basis of NIpc strands and widened to more products, beginning with essential goods, may provide an indication of the tendencies of future standards of living inside a society that rationalizes itself in the manner shown; in other words, the changes that must be implemented in the individual's behaviour not only in relation to consumption, but inevitably and as a direct consequence of all the manifestations of his social life.

Obviously, references to income are emergency references inasmuch as it would be preferable to speak of individual possibilities of acquisition, according to variably established parameters: a classic form of labour market, various incomes and profits, emoluments fixed for political reasons. In short, all kinds of purchasing possibilities of materials or services that the individual achieves in the context of the society in which he lives.

The figures I and L contain an indication of the emergent tendency toward the employment of public transport on the part of the public, in the eventuality of the costs being calculated by the adoption as of the type of analysis we have presented.

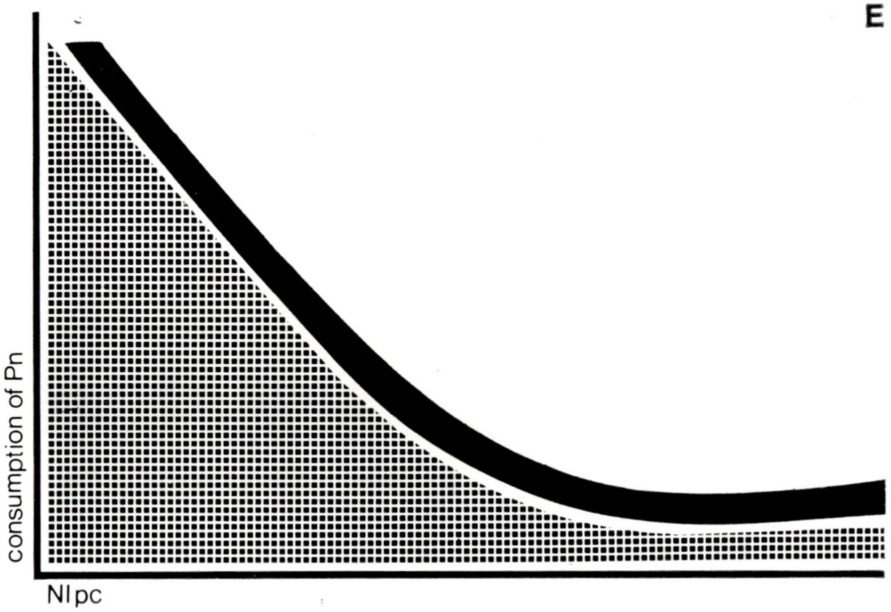

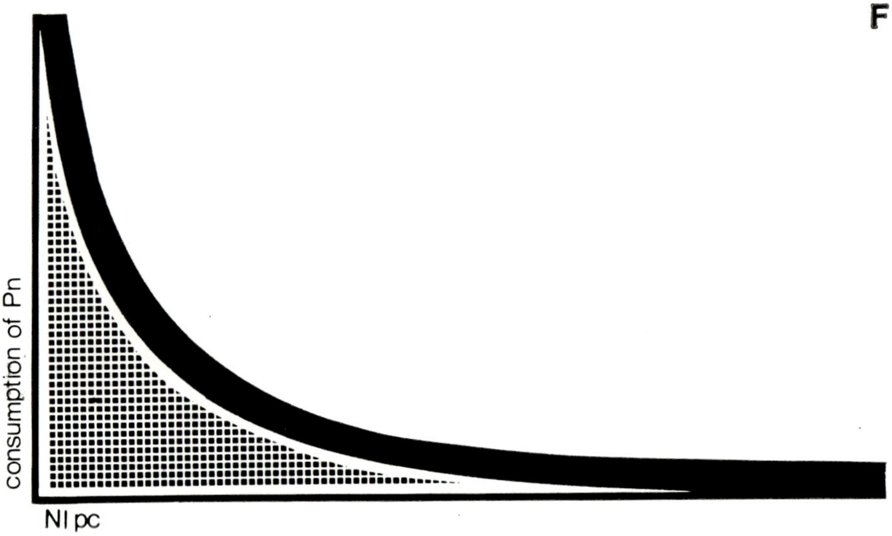

Graph E: Consumption curve for the product Pn and conventional cost expressed in income size NIpc.

Graph F: Consumption curve for the product Pn and integrated cost expressed in income size NIpc.

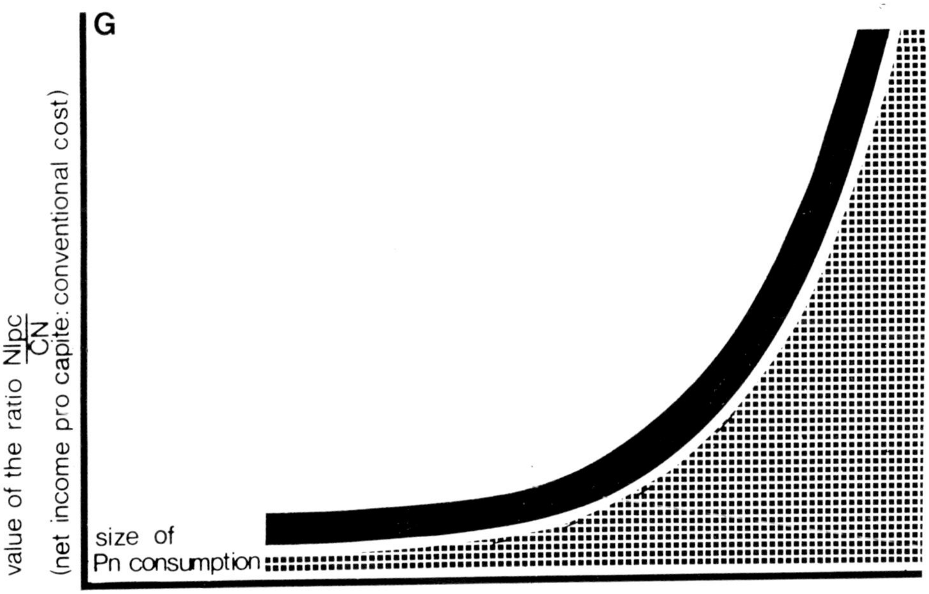

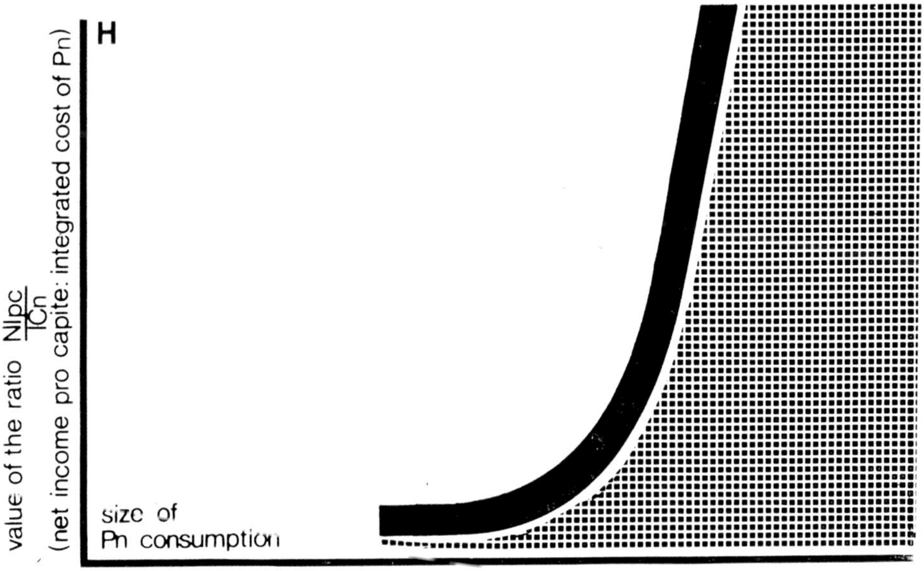

N.B.: The pattern of the two curves is also dependent on the degree of elasticity present in the consumption form.

Tendency to use public transport evaluating private transport in a conventional manner in graph I; in integrated fashion in graph L.

The preceding graphs are interpreted by taking into account that, among the many variables that influence the tendency to use public transport, the one pertaining to the net income per capita is of primary importance. This explains the pattern of the curves: an initial vertical segment (indicating the total convenience of the employment of private transport); followed by a horizontal period (indicating the total convenience of the employment of public transport); and a small and absorbable third part at a sharply angled point, between the former two. The latter is the most interesting one from the view point of the graphic construction logic. In effect, it individualizes a certain value in the national income per capita and only slightly above of it the convenience for the single recipient of income to use public transport ceases (this explains therefore the vertical segment which stands for a tendency to use them as equal to zero units) and only slightly beneath it, on the other hand, said convenience prospers and is graphically expressed as potentially infinite, precisely because of the impossibility of representing all variables (which illustrates the perfectly horizontal period in the curve). If private transport's employment costs were valued on an integrated cost yardstick, instead of a conventional one, the income per capita from which the convenience of employing private transport springs from it would prove by necessity much higher. This is represented in the second graph (L), which underlines the transposition towards higher values of the whole curve, and which contains an income strand that allows the greater - than - before and sole use of public transport.

Clearly, by following conventional cost concepts, we end up — for a host of reasons and solicitations — giving a priority to consumer products that appear (and to a certain point are) luxury goods, thus sacrificing necessary improvements in food - stuffs, not to mention the exigencies of cultural development, from the purchase of books and records to theatre attendance. There are, on the other hand, requirements such as education, participation in cultural activities, health protection, that cannot be completely satisfied as huge collective needs if the State does not intervene in a wide ranging effort to create, strenghten and renew their structures.

Let us adopt the following terminology:

Ppc = quantity of product per capita
n = number of years
R = development Rate
Pnpc = Ppc in the nth. year
PP = population
PPn = population in the nth. year
GP = global product

We notice that:

IC = (mc x Rmc + cc x Rcc + lc x Rlc) x Rpp
IC = (100 x 8% + 2 x 1 + 50 x 7%) x Rpp
thus IC = (163,5) - Rpp

After 20 years, allowing for a yearly increase of mc and 1 c of 8% and 7% respectively, and cc remaining unchanged, the value of IC will become:

$$IC = \left[100 \times \left(1 + \frac{8}{100}\right)^{20} + 2 + 50 \times \left(1 + \frac{7}{100}\right)^{20}\right] \times Rpp$$

IC = (466 + 2 + 193) x Rpp
IC = 661 x Rpp

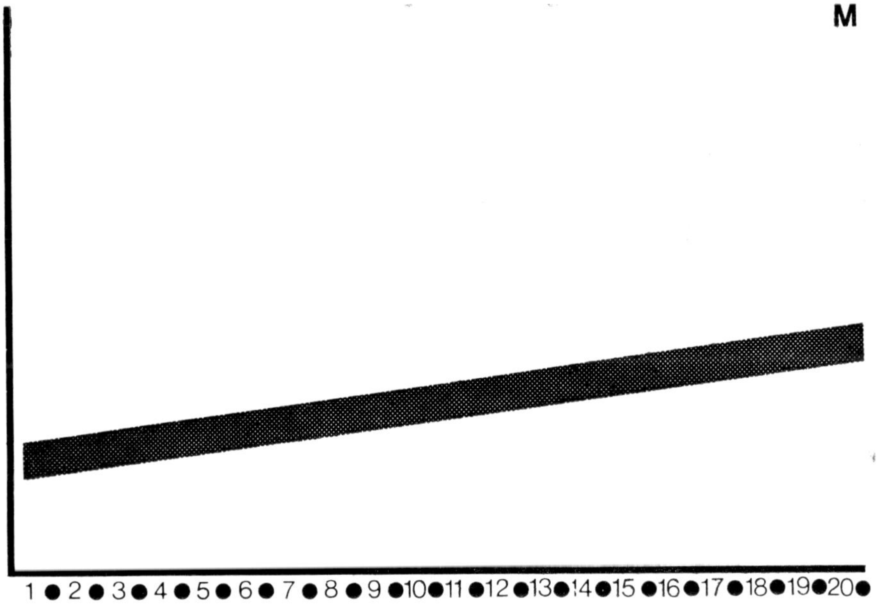

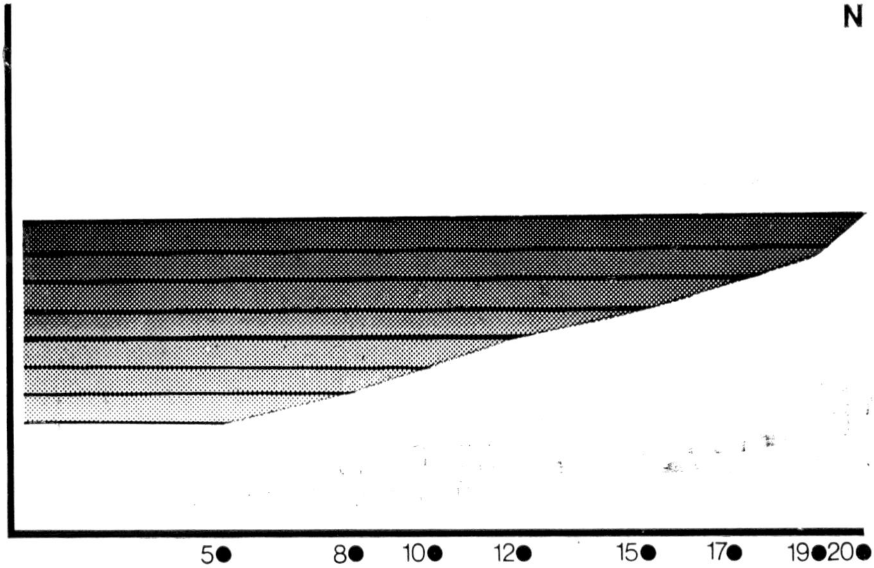

Graph M: Rm = 5% (coefficient of linear growth of mc): the curve represents the linear growth of mc per n. years.

Graph N: Rm = 5% (coefficient of exponential growth of mc; the curve represents the exponential growth of mc per n. years; mc after n. years = mc $(1 + 5)^n/100$.

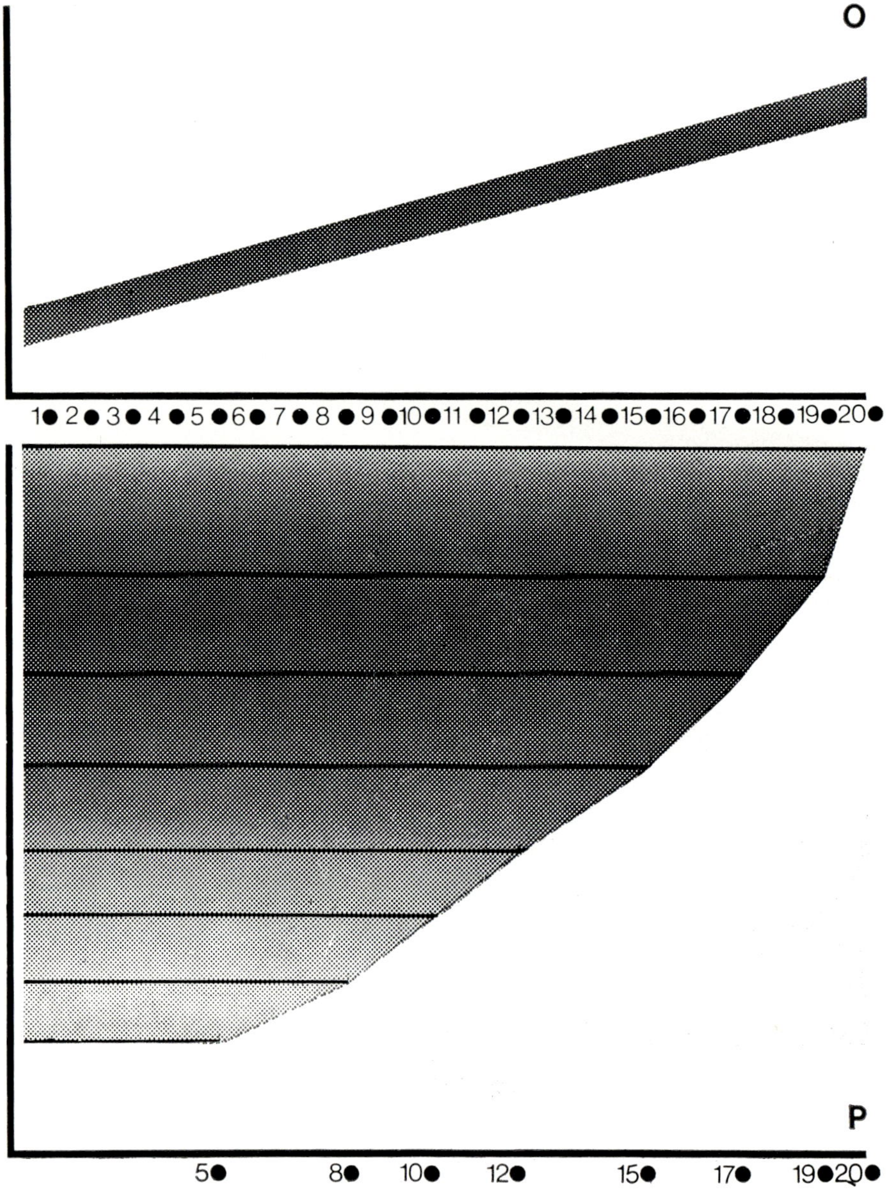

Graph O: Rmc = 10% (coefficient of linear growth of mc); the curve represents the linear growth of cm per n. years.

Graph P: Rmc = 10% (coefficient of exponential growth of mc); the curve represents the exponential growth of cm per n. years.

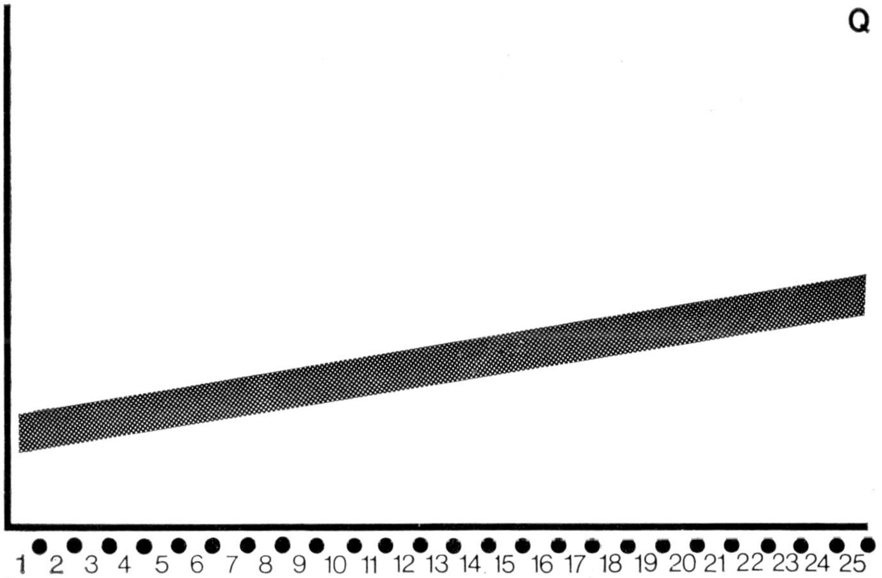

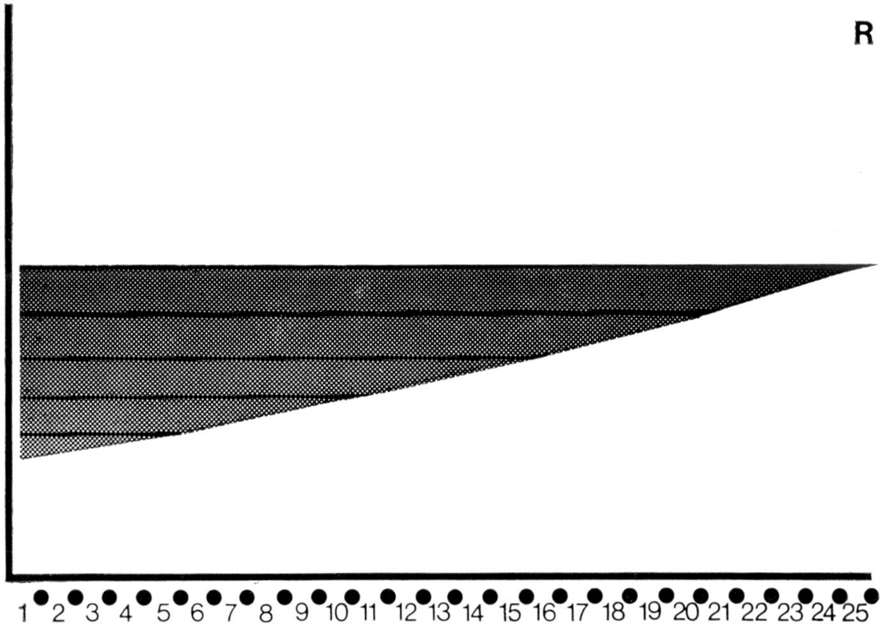

Graph Q: Rmc = 5% (coefficient of linear growth of mc); the curve represents the linear growth of mc per n. years.

Graph R: Rmc = 5% (coefficient of exponential growth of mc); the curve represents the exponential growth of mc per n. years.

S

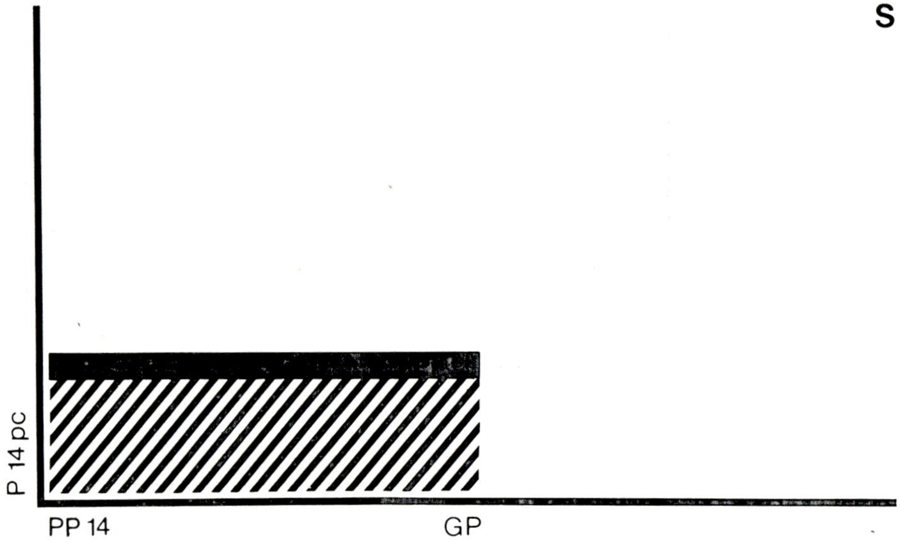

T

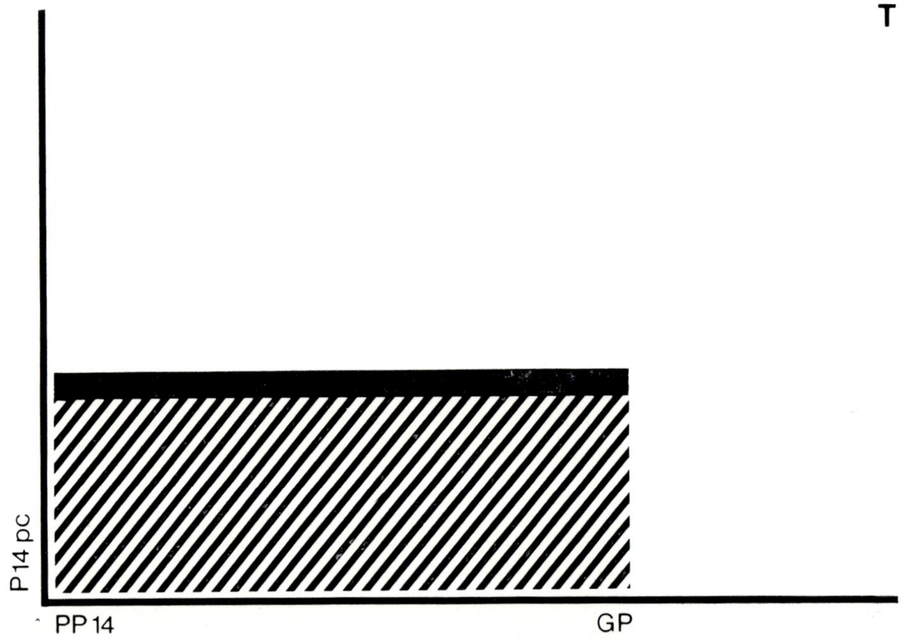

Population 10 at the beginning; allowing a yearly exponential growth at the rate of 5%.

Population consistency at the fourteenth year =

$$10\left(1+\frac{5}{100}\right)^{14} = 19{,}97993156$$

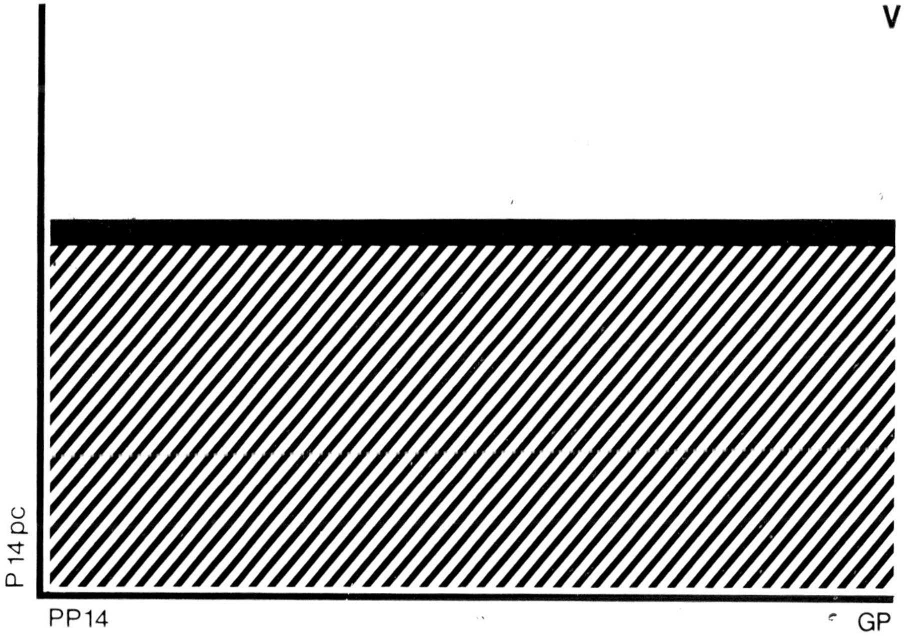

Graph. S

Allowing a yearly rate of 5% on the linear growth

Ppc = 3
n = 14
R = 5%

we have

$0,15 \times 14 = 2,1$
$3 \times 5\% = 0,15$

Pnpc = 3 + 2,1 = 5,1
PPl = 10 at the rate of linear growth of 5%
0,5 x 14 = 7

PPn = 17

Graph. T

Pnpc = Quantity industrial product P = 3 at the first year per capita; at fourteenth year allowing an exponential yearly growth at the rate of 5%

Ppc = 3
n = 14
R = 5%

$$P14pc = 3\left(1 + \frac{5}{100}\right)^{14}$$

P14pc = 3 x 1,97993156 = 5,93979468

Graph. U.

Pnpc = Industrial product P quantity = 3 at the beginning per capita; at the 14th year allowing a yearly rate of 10% on the linear growth

P14pc = 3 x 10% = 0,3
0,3 x 14 = 4,2

P14pc = 3 + 4,2 = 7

Population 10 at the beginning; allowing a yearly linear growth at the rate of 10% = 1 x 14

Population at the 14th year = 10 + 14 = 24

Graph. V

P14pc = exponential growth at the 14th year at the rate of 10%

$$PnPc = 3\left(1 + \frac{10}{100}\right)^{14}$$

PnPc = 3 x 3,797 = 11,391

Population: exponential growth at the 14th year at the rate of 10% (Population = 10)

Population at the 14th year = $10\ (1 + 10)^{14}$

10 x 3797 = 37,97

In order to construct a working model we have employed the method of system dynamics following the suggestion of the M.I.T. team which can be found in the preface to the book « The Limits of Growth ». It is stated there that the techniques of the Systems Dynamics Group has the great advantage of allowing a graphic or mathematical representation of world relationships easily understood by anyone. A perfect mastery of mathematics is not essential to understand and to contribute to the work or to use the results of such a study. Therefore anyone interested, be they students of demography or economics, or heads of state, should be able to evaluate the results and apply them with no difficulties to their field of interest.

To illustrate the structure of the working model, we wish to recall parts of the above work. It is held there that following system dynamics all systems which evolve in time can be represented in terms of two kinds of variable, characteristic and independent, which are necessary and sufficient to describe any system. Characteristic variables are those which describe the state of the system at any instant; the behaviour of the model depends on the value of each characteristic variable. Independent variables, which are determined by external agents or by the reaction of the system, determine the characteristic variables and thus control the inputs and outputs.

It is further stated there that the value of each characteristic variable is determined by the accumulation of inputs and outputs; in mathematical terminology it is the result of an integration. Inputs and outputs are controlled by independent variables which synthesize all factors (biological, political, social, variables which which contribute to modify the characteristic variable. Independent variables are in general complex expressions. In the final analysis the characteristic variables are only changed by the independent variables which in turn depend only on the state of the system through a network of *information* which is represented by influence lines and other symbols which stand for the components of an independent variable which are sufficiently important to be described separately and are therefore algebraically distinct. Influence lines join related elements, the direction of the influence beeing indicated by an arrow. The most important concept on which the structure of all systems is based is that of a *feedback loop.* Every time a change in a characteristic variable has repercussions, through a series of influences, on the starting point thus forming a closed circuit a feedback loop is established. It is wrong to think, as one usually does, of the process of cause and effect has taking place one way, and it is not sufficient to say that action A produces result B: the latter represents a new condition of the system which modifies the future of A. The loop should be followed in the direction of the arrow along influence lines, whereas the directions of the arrow elsewhere are not necessarily to be followed.

In is finally pointed out that a feedback loop can be positive or negative depending on whether a change in a characteristic variable influencing a series of elements exerts on the starting point an action which reinforces or weakens the initial change.

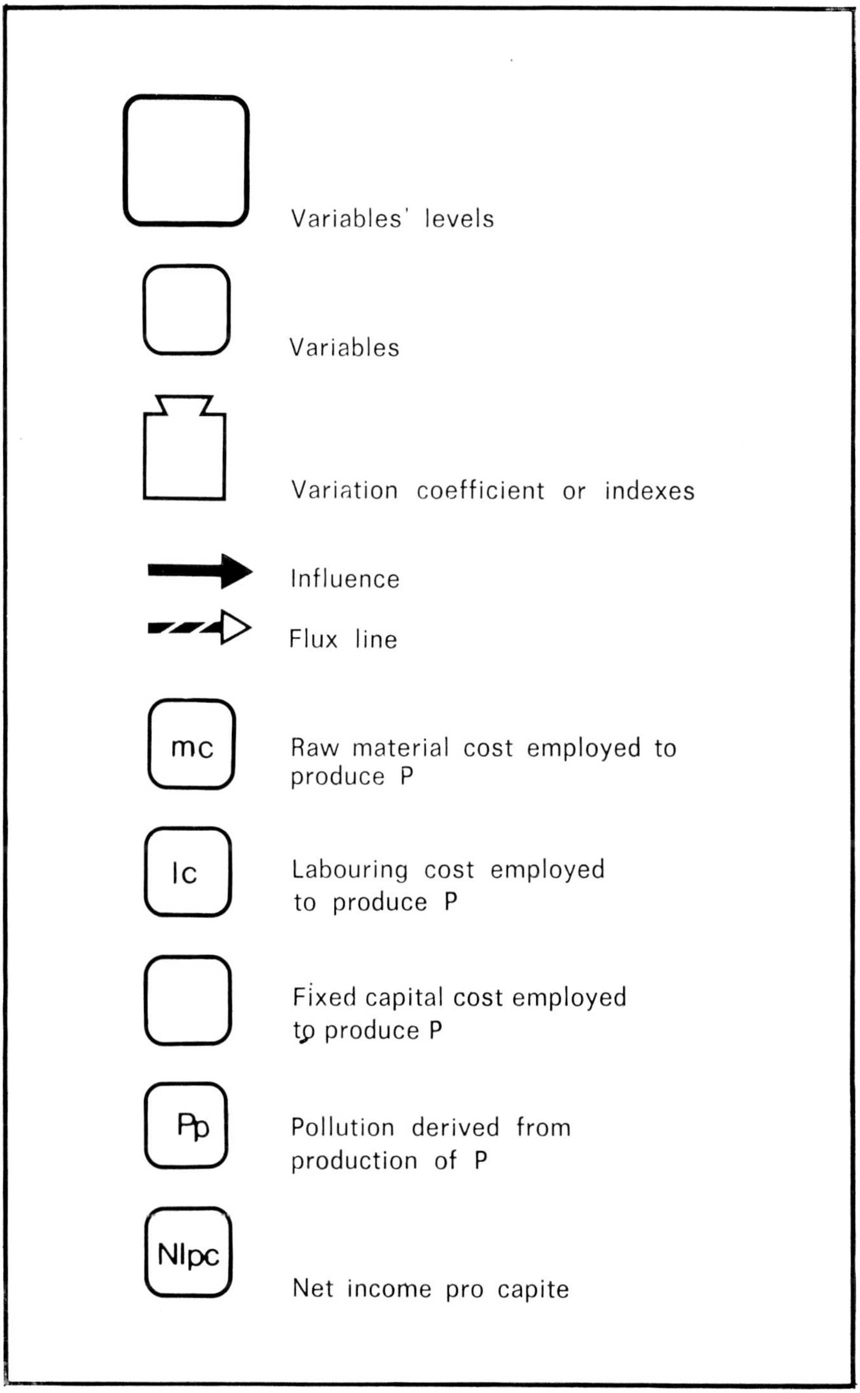
Variables' levels
Variables
Variation coefficient or indexes
Influence
Flux line
mc
Raw material cost employed to produce P
lc
Labouring cost employed to produce P
Fixed capital cost employed to produce P
Pp
Pollution derived from production of P
NIpc
Net income pro capite

IC	Integrated cost of P
M	total raw material reserves employed to produce P
L	total labour available
C	capital available
GP	global pollution
$\frac{IC}{FC}$	differentiated cost
GP	global product
CT	Category of similar products with high market possibilities
FC'	Factor of convenience for a planning of goods and services with discrimination at a categorial level
FC''	Factor convenience relative to single goods within one category
Rmc	Taxes
FC'1	Taxes
Rpp	Taxes

Rmc	Rise in the cost of raw material needed to produce P (coefficient)
Rl	Rise in the cost of labour needed to produce P
RC	Rise in the cost of fixed capital attributed to P
Rpp	Rise attributable to the cost of P to evaluate pollution damage caused by P
FC	Factor of convenience
At	Alternative technology
Re	Rise in employment
Ppc	Product pro capite

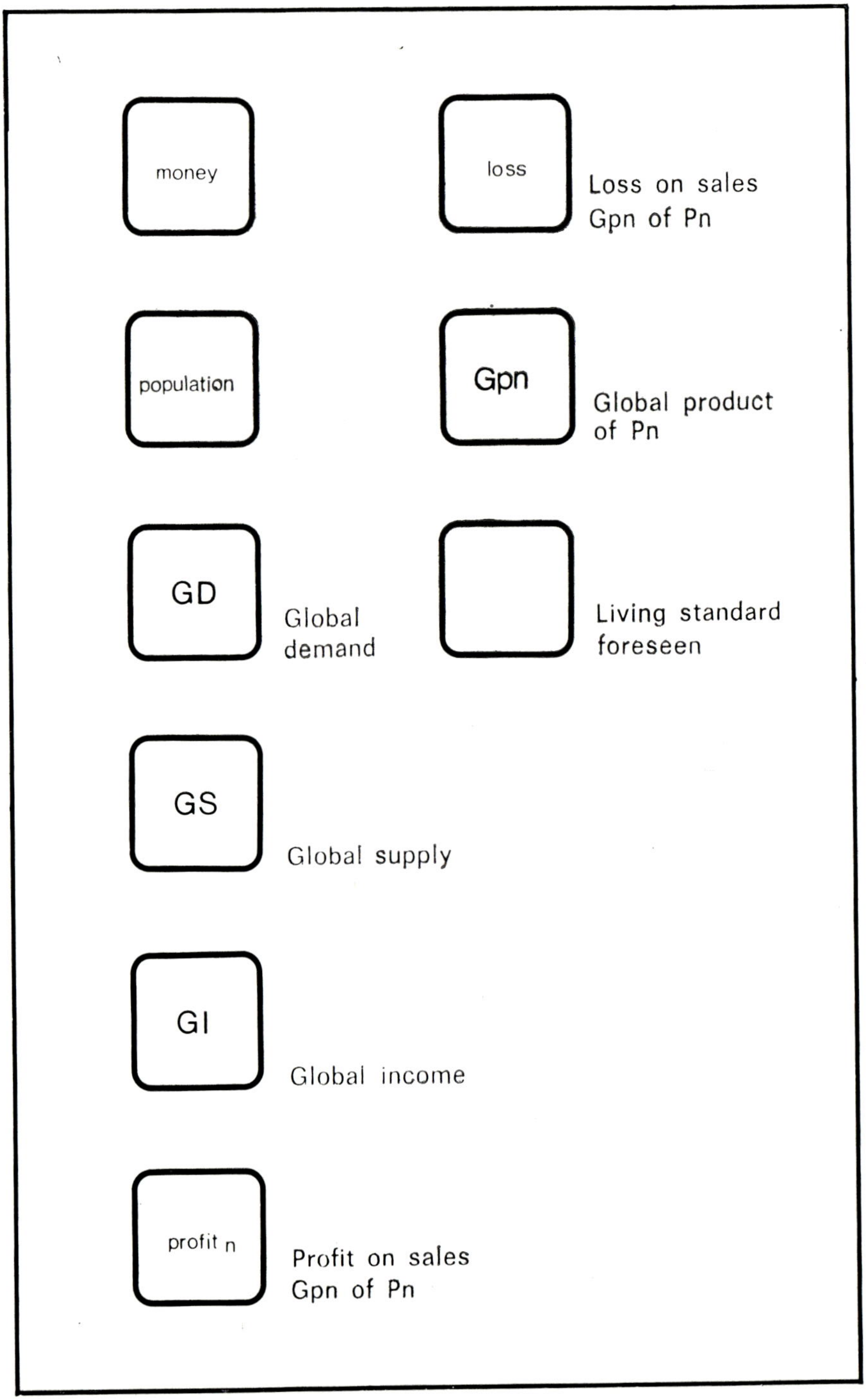
money
loss
Loss on sales Gpn of Pn
population
Gpn
Global product of Pn
GD
Global demand
Living standard foreseen
GS
Global supply
GI
Global income
profit n
Profit on sales Gpn of Pn

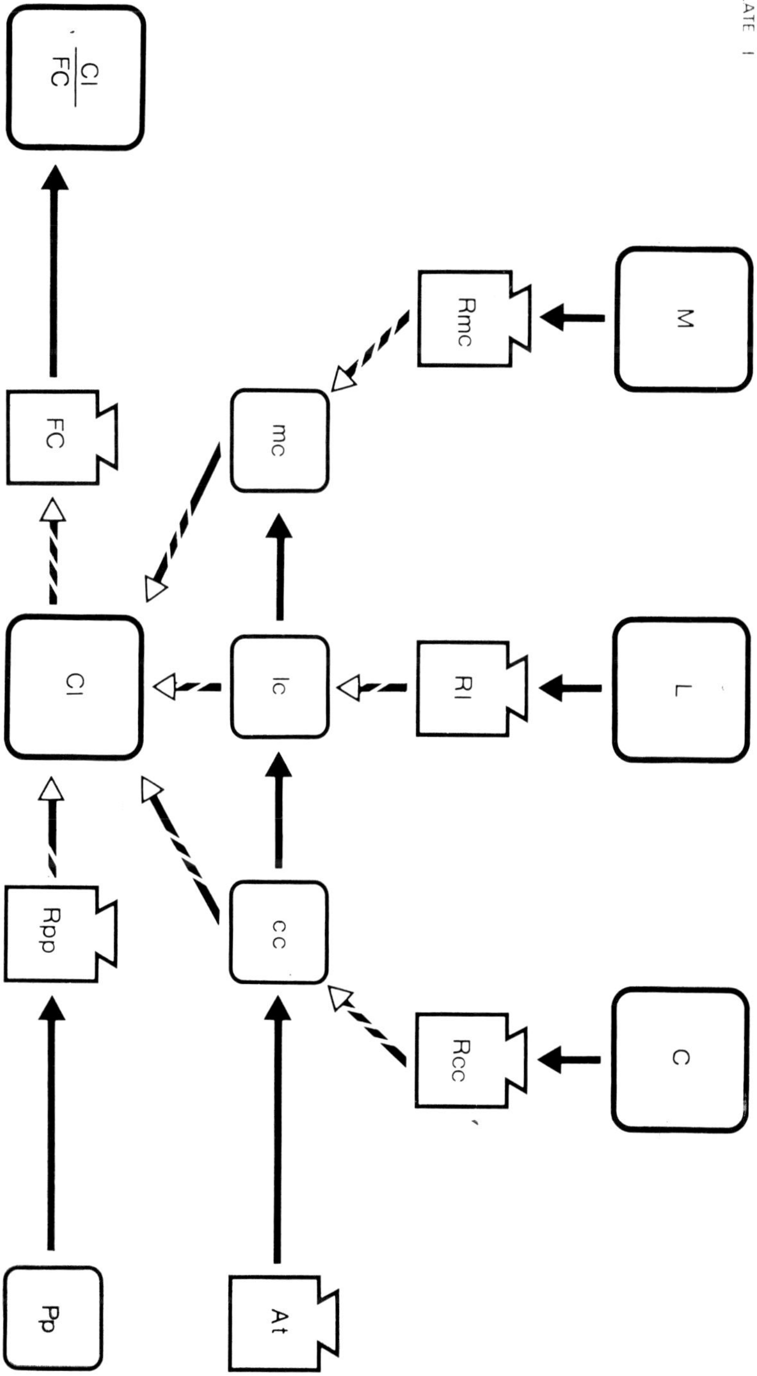

PLATE I

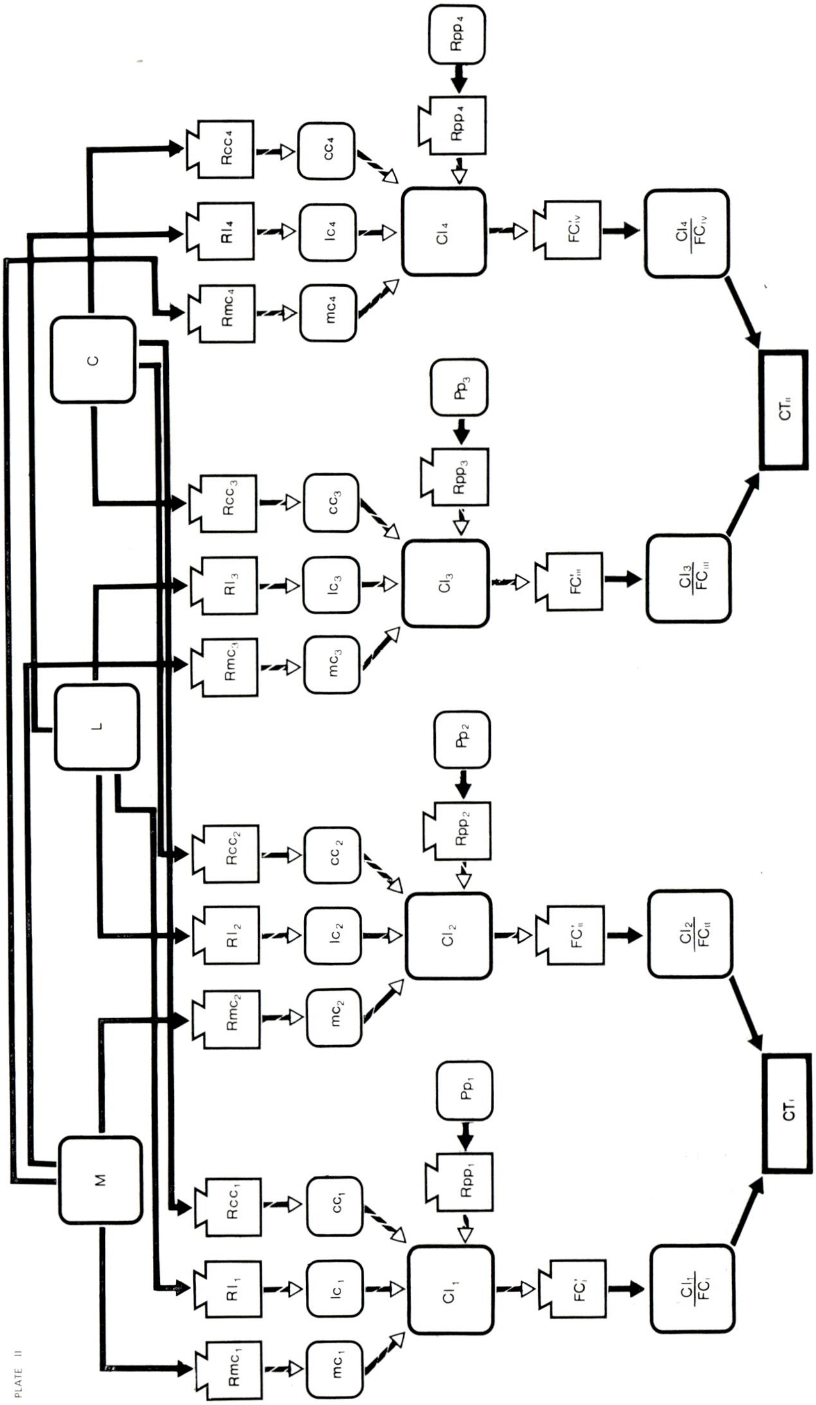
M
L
C
Rmc_1
Rl_1
Rcc_1
mc_1
lc_1
cc_1
Pp_1
Rpp_1
Cl_1
FC'_I
Cl_1/FC_I
Rmc_2
Rl_2
Rcc_2
mc_2
lc_2
cc_2
Pp_2
Rpp_2
Cl_2
FC'_{II}
Cl_2/FC_{II}
CT_I
Rmc_3
Rl_3
Rcc_3
mc_3
lc_3
cc_3
Pp_3
Rpp_3
Cl_3
FC'_{III}
Cl_3/FC_{III}
Rmc_4
Rl_4
Rcc_4
mc_4
lc_4
cc_4
Rpp_4
Rpp_4
Cl_4
FC'_{IV}
Cl_4/FC_{IV}
CT_{II}

PLATE II

LATE III

market system

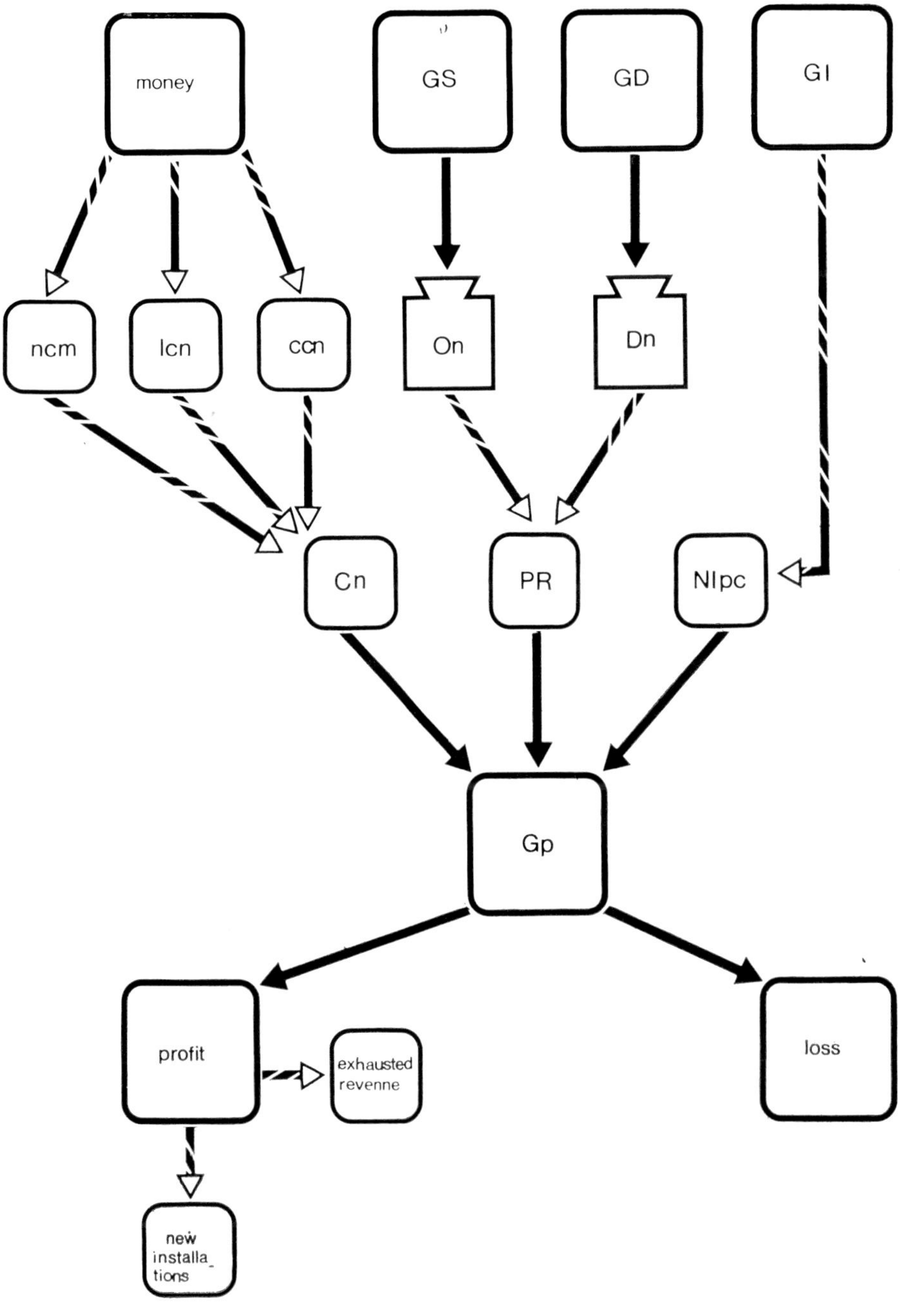

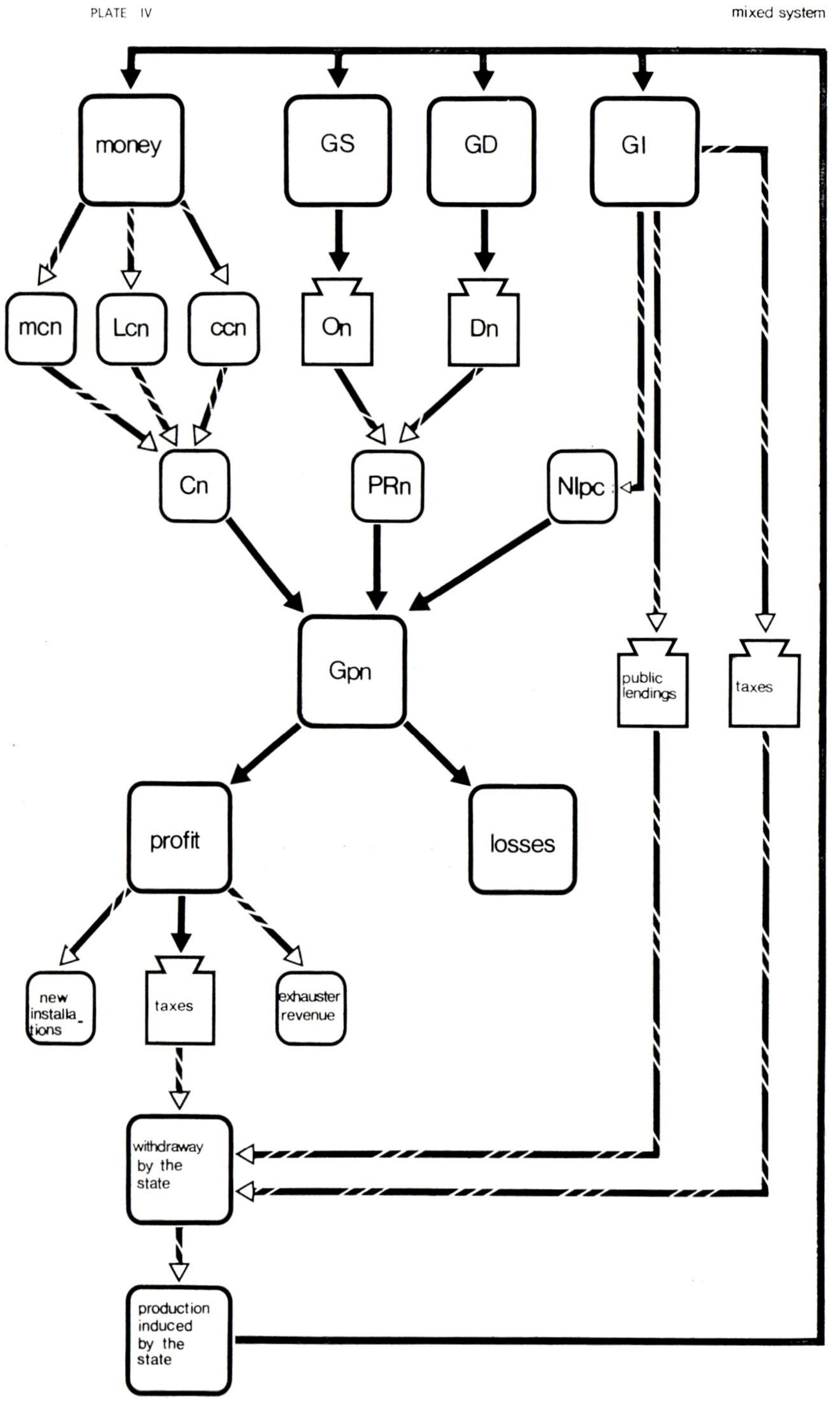
PLATE IV
mixed system
money
GS
GD
GI
mcn
Lcn
ccn
On
Dn
Cn
PRn
NIpc
Gpn
public lendings
taxes
profit
losses
new installa_ tions
taxes
exhauster revenue
withdraway by the state
production induced by the state

TRANSITION SYSTEM

The following treatment finds its roots in the fundamental premise of our research, which is that the formulation of products and services costs must be examined under the guiding light provided by new ecological knowledge and the concrete consequences of industrial development.

Aiming to create a system in which the advantages of a market economy and the necessity for programming coexist, we could attempt to intervene on the formulation of prices by transforming Rm, Rc, Ic (encountered earlier as parameters capable of redimensioning the effective incidence of raw matrials, labour, capital costs, as well as evaluating pollution expenditure) in indirect taxation through a fiscal policy geared to restrain or give incentives according to the consistency of reserves and production factors, as well the calculation of the degree of pollution produced by the various technologies adopted.

It is clear that in extreme, but by no means unique, cases the need to forbid the use of certain raw materials and unquestionably dangerous technologies must be acted upon, both because pollution will reach prohibitive levels and lethal by-products may come into existence also because of a whole series of technical considerations that cannot be analysed here.

Fiscal impositions ought to be adopted in order to fulfill, on the practical level, the priorities symbolized by our U parameter, on a categorical plane of goods (we shall emloy the symbol U') or on a single-good plane (represented by U").

Having reached this stage, the product's cost is rectified and the market may act in concordance with its habitual modalities, thus impressing on the economy the drive inherent in private initiative, the market's self-regulating efficiency, the automatism of the choices. The fiscal instrument fulfils, therefore, a social necessity in the broadest sense of the word: it safeguards the needs of an entire population, nation, state, confederation or any other kind of union.

The yield deriving from taxation should be utilized to bring the public services already in existence up to date, and for the creation of new ones; in short to create a structure that will safeguard society from the dangers that its own development and its swift, uncontrolled rise have brought to humanity for the sake of material well-being.

Having called upon fiscal instruments one may immediately raise objections as to the indirect character of said taxes, inasmuch as it runs against social considerations that ought to be seen as our true goals. Incidentally, we notice that the present use of higher indirect forms of taxations in Italy goes against the over-publicized social character of fiscal action. We may recall that IVA (V.A.T.) which legally presents itself as a tax on exchanges is nothing other than a tax on consumption suffered by people in a dearly disproportionate ratio to their income.

But, retracing our steps, it appears clear that an indirect taxation placed above the goods' retail prices will have to be rectified by a direct form of taxation apt to correct the unbalancing effects of the former fiscal intervention.

Therefore, we believe that direct taxation on individual income should be unquestionably progressive and regressive in blocks. The progressivity curve should register a moderate pattern in its initial segment, ond then soar up until it reaches the point at which an increased income becomes pointless: the aim of a truly progressive taxation. This tends to discourage great accumulations of capital; but since these are currently producing considerable concentration of capital in the hands of few individuals and private oligopolies, the aforementioned implementation of controls could only act toward the creation of a better and fairer situation.

With regard to business incomes, it is important to carry out certain observations. The already accounted-for progressivity of taxes on individual incomes would logically produce a readjustment on participation with powerful concerns. Strong shareholders would be — needless to say — negatively affected so that a wider, more popular form of share distribution may come into existence, as we have already witnessed in several countries. A reform of controls and formal verification, carried out in as objective and efficient a manner as possible, would have to be adopted.

As we have already had cause to mention, the management of large concerns ought to be conducted in a special manner by imposing on it a very particular form of controls. This with regard to truly large businesses and especially in the by-products field where a general or, essential, economic level is necessary for widespread consumption.

For large industries a comparison between costs and profits may be installed in order to localize profits not justified by either the reconstruction of obsolete installations or the necessity for scientific research. It would therefore be necessary for large industries and huge complexes to undergo a constant and purely external vigilance that, without interfering with management, should (in effect) restrict the size of factories and the quantity of financial means available, in a manner tied to the importance in business transactions and liquidity, as well as to any practical lack of an efficient competition (perhaps killed by special agreement among large producers) which would otherwise at a stroke lead to almost monopolistic situations that have the power of cutting the market's efficiency as supreme price controller.

The intervention of a supreme controlling entity capable of acting as a catalyst between living necessities and the technological progress of large firms with the consumer's interests may, then, be envisaged. It existence would have a wholly social character. Its employment would substitute the guarantee provided by competition in the realm of normal size firms with an undoubtedly efficacious institutional norm.

This kind of economic system, where the market action is integrated by a conscious evaluation of the social costs that weigh upon production of goods and services, supported by a fiscal system disincentivating (on the individual's part) the achievement of other incomes, and (on the part of large concerns) of contingency profits and monopolies, reveals the possibility of it being considered a utopia.

Its utopistic nature is as naïve as it may be to believe in the existence of:

1) an impartial and efficient fiscal organization;

2) an absolute ascertaining priority of the most important fiscal positions;

3) lack of a collision course between public powers and private enterprise, especially in the sphere of large interests;

4) controlling organizations on large firms with trained personnel able to fulfill the tasks on a technical and moral plane;

5) a technical organization that is first of all aware of the true economic situation facing the country, capable of rectifying programmes, and regulating fiscal policy according to necessities and situations as they gradually evolve. A technical organization allowed to decide which criteria to use to intervene with opportune swiftness;

6) up-to-date and fast means for the recording of facts, their transformation into objective information, bearing in mind that old information may be useful only in a comparative fashion and only to measure the true value of recent information, which must be as actual as possible.

We ought not, and indeed cannot, try to delineate beyond a mere outline a sketch of an efficient economic and fiscal organization useful for the goals prospected by us with regard to the possible coexistence of market and programmed economies. I would, however, like to clarify that what we have outlined is to be understood only in terms of a temporary phase towards a totally new system which will have to come into being through a necessary but gradual process. We shall term this a « transition » system which — it seems to us — will have to tend towards a socialist inspired economics realized through an efficient and dynamic programming.

Plate V

The symbol U may be better qualified using U' or U''. In the former instance the goods are managed by the producers according to parameters that discriminate between categories of goods. Therefore U' has the same value for all goods of one category. In the cost of the single good there is an obvious interplay concerning other factors and parameters that have variable influence on each good, even within one category. Thus U' is the only parameter (in Plate V the parameter is employed as tax) that has the same value for all goods of one category. In the latter, the goods are managed by producers under the influence of U'' (as tax in Plate V) which is different for each single good.

PLATE V

transition system

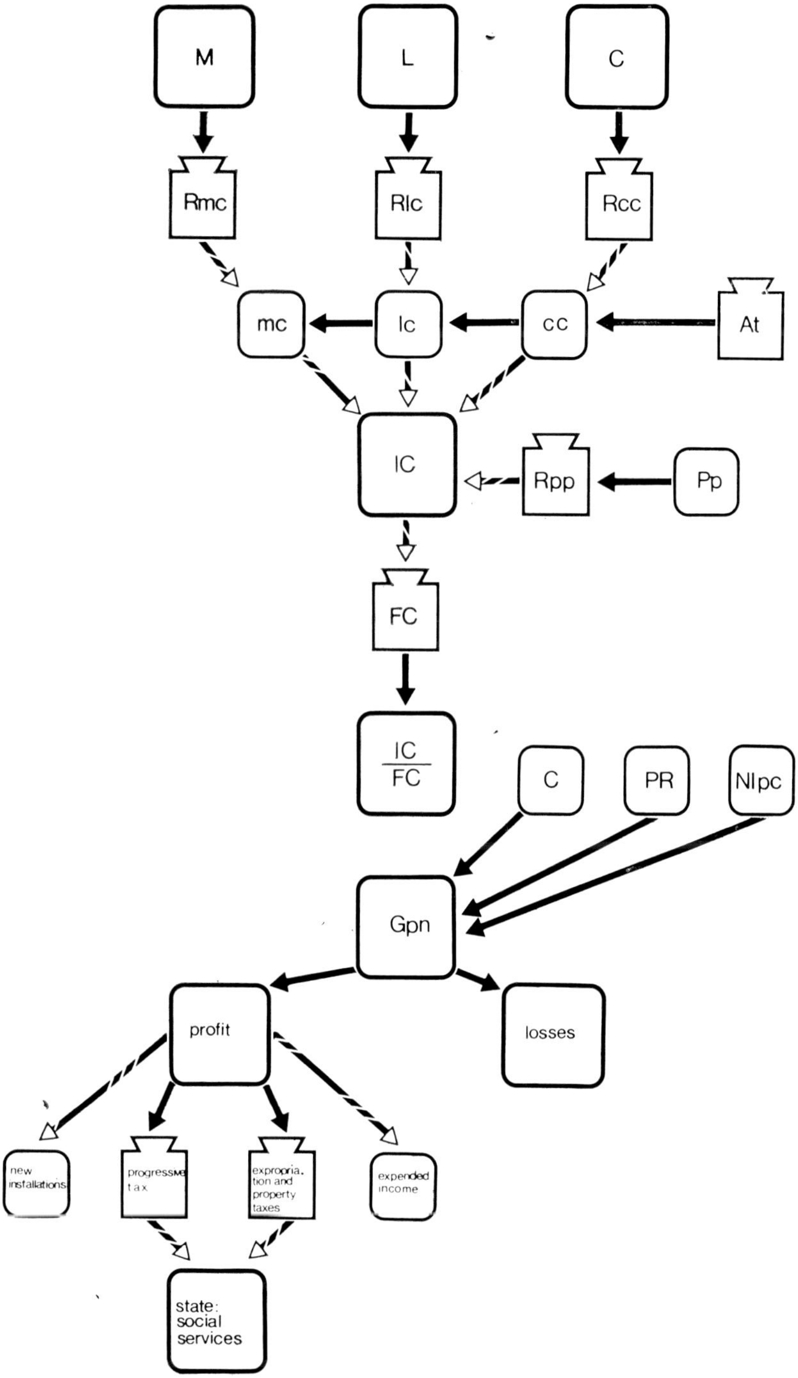

In Plate VI we find the points at which costs cease to decrease with the increase of production quantity and begin to increase in relation to added production factors for the achievement of a desired increase in production. The rise in production cost will be only temporary and then will resume a decreasing course in relation to the amount produced.

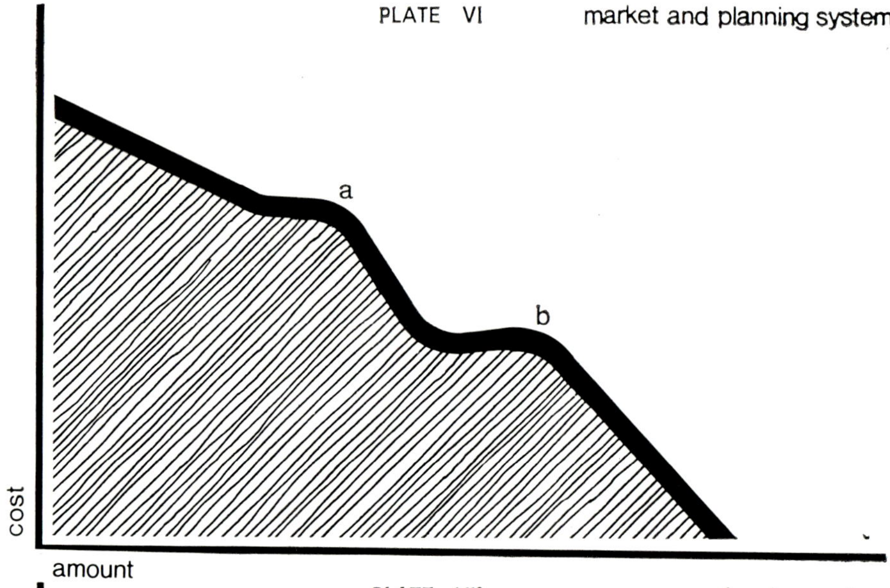

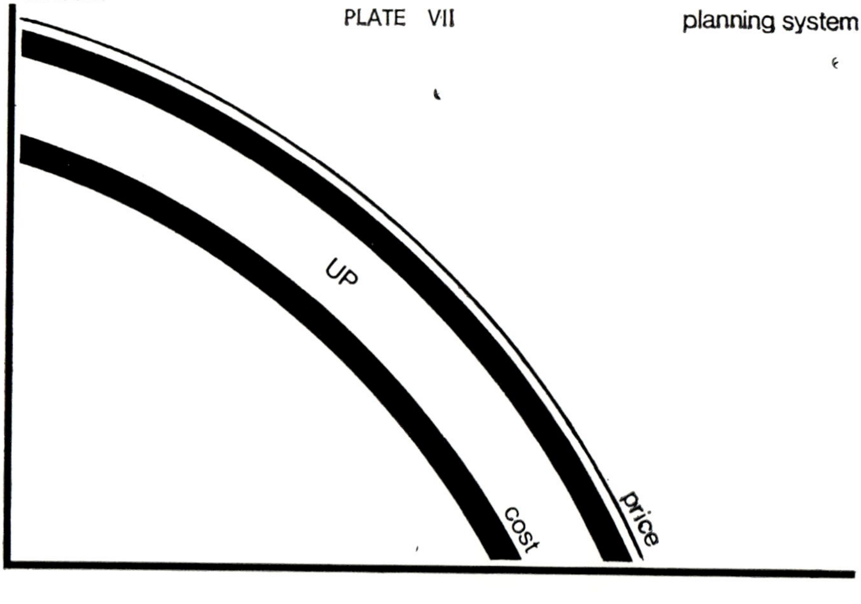

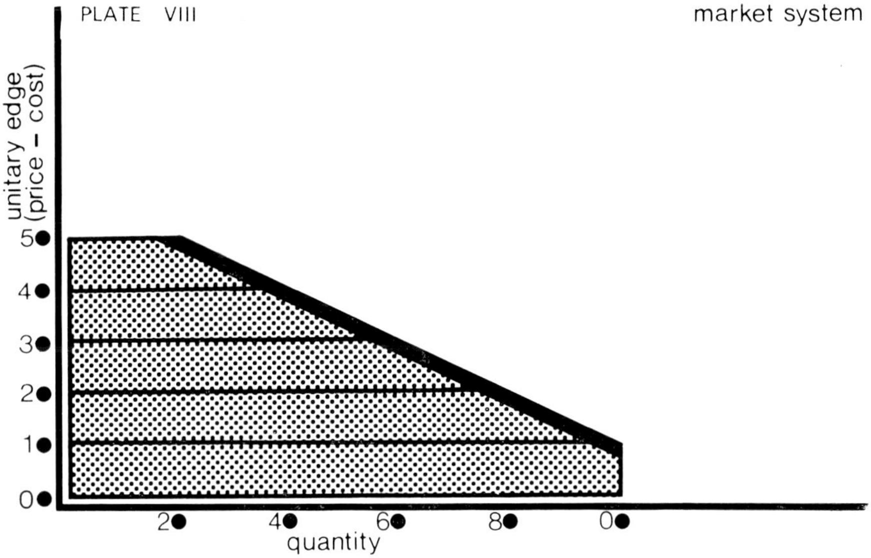

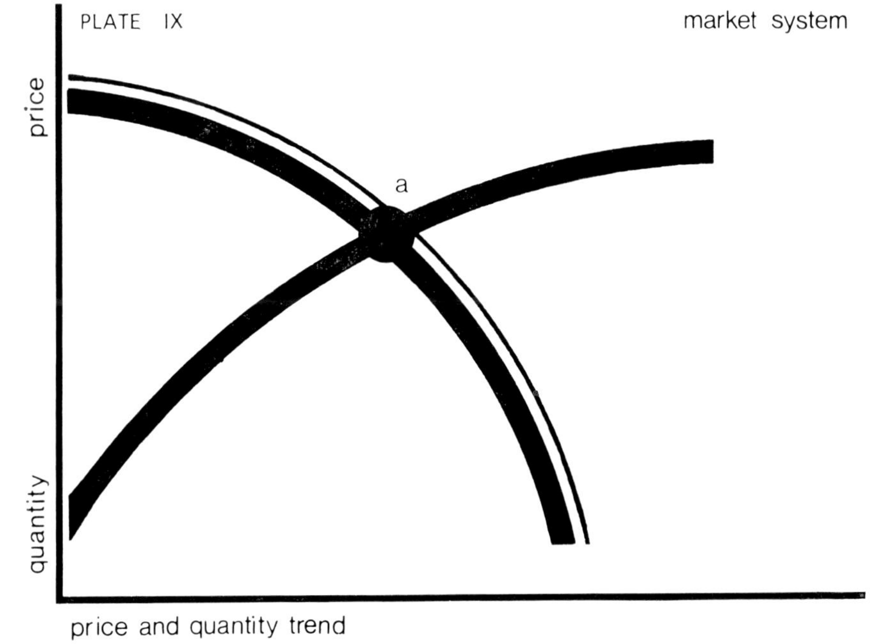

Plate IX: Superimposition of two separate diagrams used to place in evidence the meeting point between the two curves.

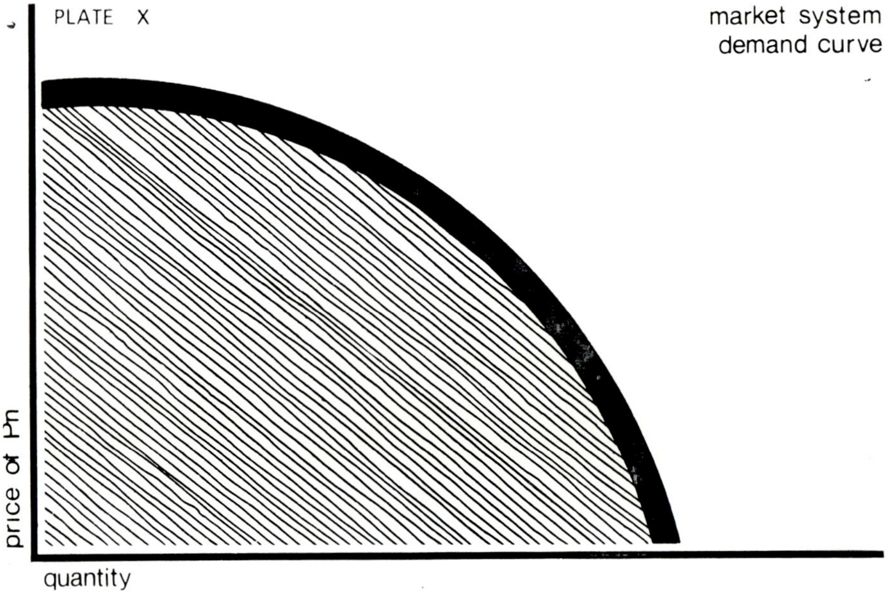

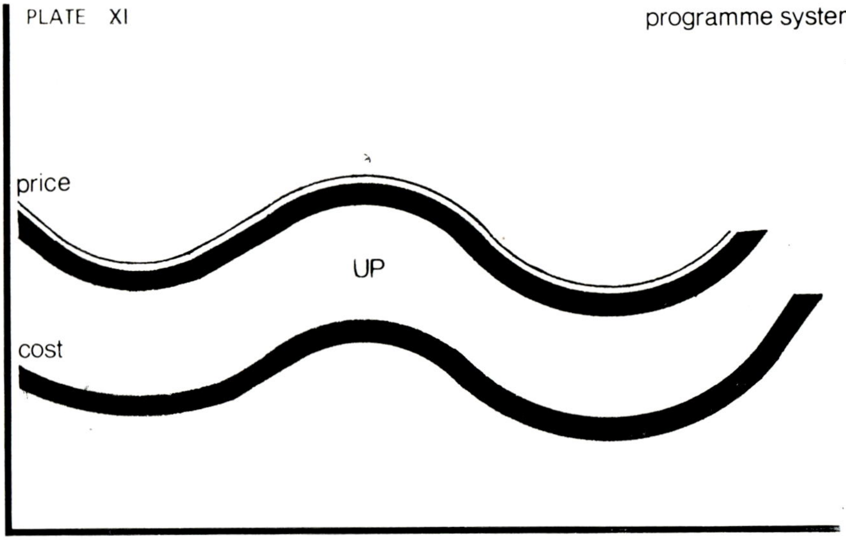

Plate XI: UP = unitary profit in a programming system.

PLATE XII

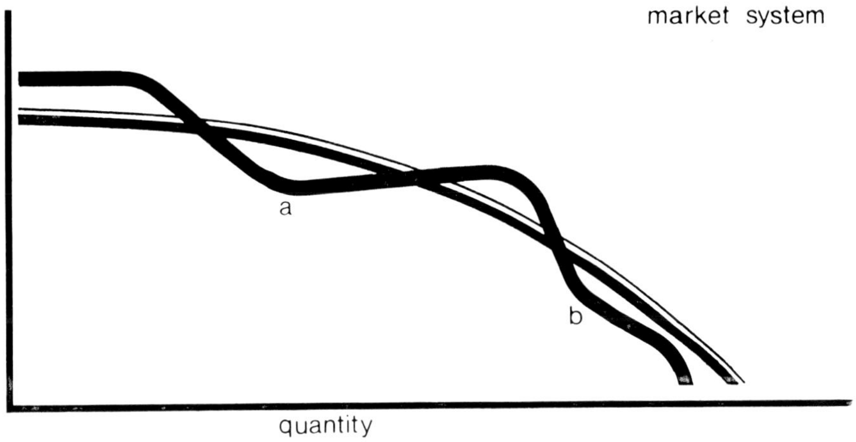

Chapter two

PROGRAMMING COSTS

DIRECT AND INDIRECT COSTS

Although the distinction between the concepts of direct and indirect costs is, on a practical and theoretical level clear, it has often been adopted by different scholars in different contexts. We have mentioned a clear distinction, but with that we do not wish to imply that to define or place one kind of cost in either category is a simple process, inasmuch as each cost may change from one category to the other whithin the dynamics of a business concern, according to the changes of conditions as may be, for example, expansion. We shall briefly examine this problem, solely from the point of view that we have hitherto adopted in our cost analysis.

In his treatise on economic logic, professor De Maria explains that the: « Direct cost is the expenditure that may ben rigorously accredited to the manufacture of a certain product, or to each unit of a certain product. Indirect cost is the expenditure which cannot rigorously be accredited to a good or a unit of that good, even if it is necessary in the production of said good ».

Having accepted that, we wish to concentrate on the concept of a contribution margin. We shall indicate with:

CM = Contribution Margin

Q = Quantity sold

RP = Retail Price

VC = Variable Costs

FC = Fixed Costs

We shall have CM = Q (RP - VC)

CM therefore gives us the global entity of the contribution that a given quantity of goods, sold in a given period, brings to the enterprise.

In other words, the Total contribution is given by

$$\text{Net Profit} - \text{variable expenses}$$

If we desire to have the percentage of contribution we shall obviously have

$$\frac{\text{Net Profit} - \text{variable expenses}}{\text{Net Profit}}$$

The above concepts are analitical praxis for the practical management of a business concern.

We have already pointed to the unavoidable difficulties that exist in the distinction between the various types of costs; they are simply imperfect classifications; but the need to acquire indexes and evaluations for concrete choices allows us to make do with inevitable approximations.

Returning on the more specific theme of our research, we would like to underline how in the application of a plan, or programme, that is a part of a transition system as the one we have outlined, the aforementioned simple relationships play a fundamental role. We shall, in fact, see how the fixed costs, in a different system from that of a free market, will be destined to be more relevant exactly because of the enormous quantity of research compared to which the effort of private industries, even the bigger ones, seems very small indeed. The problem of a greater entity in fixed costs has come into the foreground whenever we have come across the problems in public enterprise. We are therefore dealing with an aspect of the system which we hold will have to substitute in the future that of the market.

As we shall see, to approach, critically, the complexities of planning and the difficulties in concrete decisions, is misleading and unproductive. In fact, the great relevance and the dimension of the planning task will have to be faced in a condition in which the social instances and the ecological situation force unalterable and urgent decisions.

While talking about public enterprise, it has often been upheld that to the costs and the results derived from management, we must apply the same criteria which may find suitable for private enterprise. This kind of argumentation may be erroneous if the public enterprise concerned subsists in a free market system, but not in any other context.

Let us now attempt to examine, even if synthetically, where the difficulties lie, thus the costs to be incurred by the collectivity of a planning within a planning system or in the gradual phase leading up to a planning system. In practical terms, the boards and men who will have to supervise planning in their respective capacities, those who will have to approve the programmes so as to channel all efforts into an organic plan, will have to respect the fundamental clauses of the operative praxis; forecast, decision, control. Thus, posess the knoweledge to forecast, provide the forecast to act.

A man, a business concern, a state, humanity: the difference between dimensions is enourmous, but the process in man's measure remains identical. Coherence is indispensable in the various stages, but one also requires to respect the rhythms. The time for forecasting is not unlimited nor is that allowed for decisions and controls, thus, the slightest error in approximations are inevitable and must not be over estimated in the system.

In any human activity, in any form of enterprise, in any biological aspect of reality we find the concept of risk. To think that the future will make work, sufference and risk redundant is utopistic to say the least. It would, thus, be a great mistake to wish to initiate perfect plans or aim toward an illusiory perfection; most of all it is essential to be able to manage resources and deal with need in a different manner.

We must also bear in mind that forecasts are made in the light of past data and experience. Computers, with their memories, may be extremely useful to store data which enable programmers to carry out the relevant forecasts. As far as control is concerned, here too the memorization process and the selectivity of computer programmers will lighten the burden considerably.

But, in the final analysis, will the machines ever be capable of actually taking the necessary decisions? Naturally we are referring to those decisions that do not involve evaluative processes. It is, in fact, obvious that the present and foresseeable future elaborations have, and will with a greater degree of efficiency, allow them to make decisions where the evaluations are clear and quantitative, yet the problem will never be delegated to the machine where the evaluations are based on more particular estimates.

Clearly, it is difficult to find evaluative methods and criteria for certain aspects of life and human problems, to formulate objective, homogeneous and comparative judgements. Society, through evolution, will be able to modify customs and fetishes, yet it will always contain, in the spur of the moment, the motivations that determine its choices. In other words, we wish to emphasise the principle that psychological needs will always exist and are to be placed higher in the scale of priorities even to concrets material necessities. Each and every programmer will have to be aware of this.

The main hurdle in the programmer's path will be encountered when the will be faced by problems that cannot be resolved according to quantitative indications.

We wish to close our considerations on programming costs by stressing that, in our opinion, the problem does not lie in the complexity and hence the cost of elaborating an enourmous quantity of data. The planners will have to steer the ship through much worse Scyllas and Cariddis; on the one handt the danger of overestimating the number's formal exctness, on the other, the difficulty to satisfy man's non-material needs.

PROCESSING AND USES OF DATA

When dealing with programming, within a sector that is itself an aspect, or part, of a planning system, one may immediately think of statistics. Such an association may be valid and appropriate as so may be the mistrust and disillusionment of the public and technicians toward official statistics. We shall overlook all that could be said about the manipolation of statistical data for particular ends and on the errors that may be arrived at even in the most exemplary bona fide and with the most advanced statistical methods. We must on the other hand emphasize that the uses of said science cannot be but systematically adapted by exploiting the sum total of formulae and laws in the context of probable theoretic distributions, tests, discriminatory analises, decision theories, averages, prospectuses, etc, on which the free market economics has so far largely relied on.

Having stated that, we now wish to pause on more relevant concepts in management science on which even the new system will have to rely on, if nothing else as a starting point for the management of productive sectors that will come to substitute enterprise as we understand it today.

The most significant rule to be observed in this process, is a strict adherence to the concrete fact in the final checking stage. In fact, decisions are taken on a forecast bases according to the bulk of information achieved in the past, but as an index for future action, and as such, it will be affected by events that will never be identical to those that have occurred in the past.

In the light shed by this consideration, it is clear that the decision, finding its concrete expression in the near or distant future, does not require absolute precision in terms of information regarding the past.

In the final checking phase, however, absolute rigour must be applied to each control. We are not here referring to administrative or bureaucratic control, we are implying the verification of unbalanced results and the research undertaken on their possible causes. In short, what has recently been termed budgetary control, will have to assume an increasingly important role because the variables between plan and result will spring forth from much more complex sources that will require far tighter and thorough research.

Clearly, the main hurdle in the actualization of a planning system is the proper assimilation and resolution of its most diverse, hetereogeneous, elements - such as man's material and psychological needs.

Georg Cantor in his unity theory defined a unity as the fusion into a whole of objects in our intuition or in our thought that are in fact clearly different from one another. On the other hand Bourbaki claimed that a « unity » is made up of elements that are capable of interrelating with other elements within or without their own whole (cf. Bourbaki).

We believe that these considerations individualize and synthesize, from the mathematical standpoint, the fundamental obstacle in the actualization of a planned economy in man's measure.

We should mention, in conclusion, that for the purposes of control, the establishment of a cost standard, is more appropriate than a classification of costs, even if we will have to follow different criteria from those inherent in the capitalist economy.

Clearly, a planned economy can only be put into practice if the population quota and the territory it refers to are sufficiently vast (thus space is an element in the problem). Programming will have to take into account the condition of the land, the geographical location, the environmental situation; in short the ecosystem within which the plan is supposted to take shape.

A revaluation of spatial or dimensional economy in relation to a puntiform economy cannot but acquire great value in a planning system, since it does not require such accuracy in the references that imply flexibility of action rather than the rigid application of economic concepts.

An examination of the importance assumed by space in the economic thought from the sixteenth to the eighteenth century (cf. Pierre Dockes) would reveal that the integration of space, like that of time, is absolutely necessary for an economic theory. Most contemporary economists are rediscovering yesterday's problems and consequently they merely reinstate solutions that have already been perceived in the past. The concepts as well as the problems treated by the « classical » authors we have already mentioned thus find particular relevance today.

Chapter three

MANAGEMENT TRENDS

OPTIMAL RELATIONSHIPS BETWEEN COST AND PRODUCTION

On the basis of the environmental considerations thus far developed, it is clear that the management of enterprise in a planning system, which has been rendered necessary by the development crisis, will have well defined characteristics. The objective circumstances of greatest relevance to our research are: an increase in direct costs (which we already mentioned in relation to the integrated costs concept); an increase in indirect costs due to planning expenses; restraint imposed on the size of the enterprise determined by the concern's location.

Continuing along these lines, would lead us in a different direction from that we intend to follow. We shall therefore limit our observations to those that are exclusively directed and relevant to the organization and management of the enterprise in the context we have hypothesized. Let us therefore move on to expose general criteria on arguments pertaining to a complementary future reality.

We regard as primary objective the modification and the adaptation of our present operational technology based on the concerns' own efficiency. More explicitely, we could indicate our objectives as being the attainment of « an optimal relationship » between the volume of production and the costs implied by it.

Before considering the choices, the precautions and the efforts to be made in this direction, we need to realize that in principle, the politics of an enterprise change according to whether direct costs soar or slump. We have already seen that in our case direct costs can be considered as consistently soaring. In fact resources are constantly dwindling both as far as raw materials and energy are concerned. The exploitation costs for energy and raw materials are destined to increase as always poorer ores are utilized in a situation of advanced exploitation.

Even agricultural costs are destined to increase while expanding the more barren areas. From this follows an increase in the price of all agricultural products and the means employed or those relying in their production, from natural textile fibers, and zootechnical resources to various vegetable provisions and non synthesized chemical products. The obvious consequence of such a state of affairs is that the tax margin is considerably reduced. It is imperative therefore to determine very accurately the point at which the expenses incurred in boosting production cease to yield a positive result and begin to produce an economically negative one.

Even if in a different situation marginal production may be pursued, in the situation we have outlined, an increase in production poses the problem of achieving it without negatively burdening the management, thus the general criteria adopted under the stimulus of demand on the free market are no longer valid because the eventual increase in the price fetched often come up against higher costs than those of marginal production.

In a planning economy, which we believe represents the only viable solution to the very dire problems caused by the radicalization of the environmental situation, enterprises will have to adopt managing criteria which imply a more accurate analysis of costs and operations. This analysis should not only judge the global result, as we are generally inclined to do in favourable situations, for a context which does not require excessive caution, also because the interest of the community is not an element in the profit logic. Even today and even if one cannot talk about an authentic free market, the logic of enterprise remains always the traditional one, with complete disregard for the community's interests. This is naturally not a moral judgement, but a simple statement of fact in an irresolvable situation until enterprise is not framed in a planning situation where the general social interests carry sufficient weight and value. The analysis we propose for management enterprises is that which proceeds in examining situation by situation, case by case, initiative by initiative. In other words, the operations of each enterprise must be judged in the cold light of the costs which it creates or of the resulting achievements. We cannot limit ourselves to simply observe that a set of means or operations produce a global result that is economically positive. To be satisfyied with averages of final results as a mean of managing an enterprise is to ignore that, very probably, a part of the profitable operations pays the expense of the unprofitable ones. Reducing one's contribution and therefore curtailing the margin destined to cover the indirect costs which must increase due to the heavy onus of programming, the global result can become negative even if within the set of operations which constitute the management of an enterprise several are positive.

In short, one must set aside the healthy part of the management and resolve in other ways the situation for which normal solutions have proved too expensive. We wish to stress that bearing in mind the community's requirements, the logic of an enterprise remains rigid even, and above all, within the planning system. Here miscalculations are more serious exactly because the enterprise must fulfill a new social function. Where in the free market, enterprise may justify itself at the expense of other interests, within the system it must respect a more rigorous logic since it is pledged to a greater number of commitments and requirements.

Another observation is that the financial side cannot certainly be neglected in a planning economy. On this score, in the proper management of an enterprise, we wish to underline the necessity of carefully evaluating the possibility and suitability of new installations. Often the redemption of certain expenses that appear on paper have already been invested under other headingds. It is therefore necessary to verify this in evaluating the financial obligations that have to be met. If a growing enterprise has previously made new commitments the funds are no longer available to renew the installations for which they had been allocated. Wanting to increase by one unit the enterprise's production, one must verify that the increase of production by one unit manifests itself in an increase in revenue that exceeds the crescendo in costs arising from intensified production. It is too easy to misinterpret this last point. All the implications of one unit of production must be foreseen. To determine whether it ought to be included in the enterprise's framework it is not sufficient to ensure that its contribution is greater than the direct costs; if we follow such a criteria we shall be fatally led to include gradually new units thus creating costs that are known to be fixed. We would have control over the enterprise's dynamics; the set of units produced by an enterprise must therefore be determined in the light of a highly discriminating analysis.

The entity problem must be tackled correlating direct and indirect costs and verifying the validity of the business according to the overall performance of the enterprise. The appropiate size, however, is also determined by the variety of the articles produced and finds its optimal efficiency in the numbers and the combinations of articles which best reflect the potential and the technical characteristics of the installations. Above or below certain quantitative limits and certain combinations of articles produced, the efficiency decreases and only through a change in the size and characteristics is it possible to raise it again. In a planning economy and in a difficult situation, the choices of articles to be produced must be made on the basis of research and considerations that we have already pointed out at several stages in our studies.

We have mentioned a parameter which we indicate as « FC » which stands for factor of convenience. We will have cause to mention it again and in particular in relation to « values » investigating the constant term-man-as a parameter and partaker in the socio-economical solutions, undoubtedly, the diversification of articles can only be reduced when compared with the present situation. On the other hand, the plethora of products, the huge quantity of manufactured goods, will present a cost which will, fortunately, be avoidable while promoting an analogous product. The savings in the diversification of products and manufactured goods will add to the expense of goods and services introduced for the common use without real need. These are requirements that have been introduced artificially and their elimination or reduction will prove to everyone's advantage.

It is impossible to talk about management without mentioning the organization of the enterprise's activities. This aspect will assume a growing importance in enterprise within a planning system framework. The development of a plan, the refinement of data processing, the interpretation of statistical data, the elaboration of information, the anticipation of management results, the formalization of enterprise procedures are but a few of the responsibilities resting on the organizational directive activity. It is especially necessary in a planning system that these tasks be given to minds that have direct knowledge of the matters involved and implied.

If in an enterprise that operates a planning system the managers should have been formed within the framework of that theory and managing techniques that have influenced professional management courses, the results would be totally disastrous. The findings of a whole school of thought on the scientific organization of enterprise has already been placed in serious doubt, but we shall pause on the reasons that warn against employing in managing positions, elements that have exclusively been the recipients of a theoretical formation. It is inevitable that this type of manager should feel greatly confused by the difference between the real world and its expression on paper. Whoever has not and does not live within the organism of an operating enterprise will tend to give greater emphasis to fewer numerical data, and will be liable to proceed on the basis of a schematization of the enterprise's facts which has little to do with the concrete reality of these facts.

It is often very sad to witness that people who possess a wider and thorough grasp of business knowledge and practical market conditions, should be ostrichised merely because they do not have the ability to theorize according to the procedure in fashion. In other words, one ought to listen more closely to those who know the facts and the reality they refers to better, while those who are destined to direct and to plan should be placed in contact with the practical development of the problems.

Endless mistakes and misunderstandings are determined by the «air cushions» on which some managers move and from their aversion to even discuss matters with the people that have direct information on the facts. This has resulted in badly collected data, interpreted without sufficient critical skill to notice incongrueties, unjustified conclusions that may only lead to rash decisions. But the worse aspect of it all is that in the acquisition of data or in the realization of given procedures these people are influenced by the illusionary research of a formal perfection without considering the costs, the real availability of the necessary means not to reach but even approach such formal perfection. The dynamics of the reality of enterprise is such that overly detailed configurations and representations are bound to become obsolete in a very short space of time, thus, aiming for a less detailed but more up to date overview of the reality of enterprise is undoubtedly more fruitful in the long run. Lastly, one must be persuaded that, to use biological terms, the enterprise possesses a metabolism of its own that cannot be disregarded, so that one can never overlook the size, the capacity, the means on which it may rely in its effort to reach the enterprises' goals, not with theoretical assets but with the concrete assets available to her.

On this bases it is clear that the problem of the enterprises' organigram should be examined from a horizontal and no longer from a pyramidal standpoint. In practice, each sector must have its responsible head that operates directly on each matter. Such a group leader must possess a first hand knowledge of the problems involved by personal experience without mediation. He must have ample responsibility and the power to act within the framework of predisposed plans, thus finding himself in an autonomous position on an operational level and within the boundaries deemed fit in advance. Thus the much deplauded and unfortunately far too common situation where responsibility is practically non existent inside a system that tends to organize itself vertically (there we find a pyramidical hierarchy where responsibilities are divided and confused between head and assistants, so that it becomes virtually impossible to pin down particular mistakes of oversights on any one individual) may be avoided.

Generally, in an organization of a certain size, those in charge are taken to justify those below them (even for purely personal reasons) all along the hierarchical line, with the result that the situation becomes rapidly more confused.

With separate and indipendent sections, the comprehension of the enterprise, or a group of enterprises', workings becomes clearer. In this context, of course, a manager's function varies according to the responsibilities alloted to the heads of the various departments. Having said this, it is necessary that between the source of decision and the leaders of the different operative groups there be as few diaphrams as possible. These diaphrams are often the sources of rivalries, misunderstandings, jealousies that create problems which have nothing to do with the business concern efficiency and which actually reduce it.

It is therefore important to decentralise operational activities to the departmental heads, dividing decision-making within the preorganized plan, without burdening a manager with all decision making since that would lead to the error of overloading a single individual.

The need for coordinators will be recognized and the control and verification of the realization of the plan in the operation will be delegated to them. These coordinators will also have to be concerned with the complex of services, maintenance and supplies and the relation of the production sectors with the various types of controls and verifications; foreseen in such a way as to ensure the smooth functioning of all subsidiary sections.

The organigram will also anticipate tasks which tend to eliminate the frictions and antagonisms that often develop within enterprises.

Psychologists have already found employment in many business concerns. In a system such as the one we have outlined, psychological assistance will have different objectives and the problem of motivation will be otherwise examined.

In conclusion, we may affirm that we are confronted with a well arranged organization when the following conditions are present; the objectives of the enterprise are clearly defined, the plans to realize these objectives are efficaciously described; a detailed distributions of tasks is undertaken; clear working rules are stated.

CHAPTER FOUR

PLANNING AND COSTS OF SOME SERVICES

DISTRIBUTION

The question of distribution is of interest to us here because the problem survives in all its complexity also in a planned economy or one which leans toward it.

A prerequisite to the planning of the points of distribution is the investigation on the size and costs of the point of sale. It is evident, therefore, that a point of sale must have an income in order not load too heavily the products presented to the public with distribution costs.

We state the problem concretely in order to indicate the method through which a given result may be obtained. In this analysis we consider three kinds of costs: 1st) cost of the product sold (we include here all costs which may be considered *direct* according to the criteria already elaborated, for example the expense incurred in transportation of the merchandise from point of production to point of sale); 2nd) direct costs of each point of sale; 3rd) indirect costs made up by the organization of the global distribution system, including expenses incurred in planning, research etc.

INDIRECT COSTS AND REVENUES OF THE POINTS OF DISTRIBUTION

We proceed to the examination of the following schemes:

Scheme I — Finding the number of points of distribution, all other management issues being known.

We indicate by IDC the indirect costs of the entire distribution system;

We indicate by RU the average contribution per unit of product sold;

We indicate by DPC the direct costs of each point of distribution;

We indicate by P — management profit;

We indicate by S — total sales (number of units of the product sold).

We have:

$$\frac{IDC + P}{RU} = y$$ = number of units of the product sold to cover all costs not arising from the point of distribution or not included in the calculated average contribution per unit of product sold.

$$\frac{DPC}{RU} = x$$ = number of units of product sold to cover direct expenses of the points of distribution.

$$\frac{S - y}{x} = NDP$$ = number of points of distribution compatible with the expected management result.

$$\frac{S}{NDP} = RDP$$ = average revenue of the point of distribution.

Whenever the indirect costs of organizing the entire distribution system can be reduced through an increase in efficiency, there is the possibility of increasing the number of points of distribution available to the public. In other words, it is possible to allow points of distribution with a revenue lower than previously correlated with the fall in the indirect costs of the over all organization.

Scheme II — Finding the correlation between the fall in indirect costs and the theoretical number of points of distribution.

Let AF be the amount by which indirect cost fall:

$$\frac{IDC - AF}{RU} = y$$ = number of units of the product sold to cover all costs not tied to the point of distribution or not included in assessing the average contribution per unit of product sold;

$$\frac{DPC}{RU} = x$$ = number of units of product sold to cover direct expenses of the points of production;

$$\frac{S - y}{x} = NDP$$ = number of points of distribution compatible with the expected management results, after the fall in indirect costs afforded by the management savings;

$$\frac{S}{NDP} = RDP$$ = average revenue of the points of distribution, after the expected fall in indirect costs.

Scheme III — Predicting a summary of correlations between AF, NDP, RDP supposing AF 1< AF 2< AF 3 < AF 4 and S constant.

S	AF 1	NDP 1	RDP 1
S	AF 2	NDP 2	RDP 2
S	AF 3	NDP 3	RDP 3
S	AF 4	NDP 4	RDP 4

the NDP s will take on values:

NDP 1 < NDP 2 < NDP 3 < NDP 4
RDP 1 > RDP 2 > RDP 3 > RDP 4

The indications of scheme three are a consequence of the other two schemes.

The same will hold for scheme four.

Scheme IV — Predicting a summary of correlations between S, NDP, RDP, supposing S 1 > S 2 > S 3 > S 4 and the AF constant.

S 1	AF	NDI 1	RDP 1
S 2	AF	NDI 2	RDP 2
S 3	AF	NDI 3	RDP 3
S 4	AF	NDI 4	RDP 4

The values of NDP 1 will be:

NDP 1 > NDP 2 > NDP 3 > NDP 4

RDP 1 < RDP 2 < RDP 3 < RDP 4

From the symbols and the formulas exposed, it follows that the higher the indirect costs of the organization as a whole, the higher must be the revenues of the points of distribution. Of course, it is necessary to control the existing points of distribution, which implies increased inconvenience for the public. Conversely, the cheaper the planning, the supervision and the control, the more it will be possible to refurbish the field of points of sales taking care of marginal needs, which are actually the most expensive. Ultimately the savings achieved at the vertex gives rise to the possibility of affording a greater marginal cost in order to better satisfy the needs of the public.

All this lies on the hypothesis of a constant overall cost, obtained by adding the cost of supervision to that of the individual points of sale; this shows that the better the central efficiency, the greater will be the advantages presented to the public under the form of either a greater number of points of distribution (convenience) or reduced cost of distributed goods.

Following scheme III, we have developed a mathematical example quantifying a series of values for AF and holding S fixed; Nm and rm are varied corrispondingly. Plotting the values of Nm and rm on Cartesian axes we have traced a curve.

From scheme IV we have obtained in the same way a curve varying S and holding AF constant.

Here follow the tables and the curves:

AF =	0	= NDP =	26,5	RDP =	63,010
AF =	4.815.808	= NDP =	27,5	RDP =	60,728
AF =	9.631.616	= NDP =	28,5	RDP =	58,596
AF =	14.447.424	= NDP =	29,5	RDP =	56,610
AF =	19.263.323	= NDP =	30,5	RDP =	54,754
AF =	24.079.040	= NDP =	31,5	RDP =	53,016
AF =	28.894.848	= NDP =	32,5	RDP =	51,384
AF =	33.710.640	= NDP =	33,5	RDP =	49,851
AF =	38.526.464	= NDP =	34,5	RDP =	48,405
AF =	43.342.272	= NDP =	35,5	RDP =	47,042
AF =	45.158.080	= NDP =	36,5	RDP =	47,753

AF = 52.973.888 = NDP = 37,5 RDP = 44,533
AF = 57.789.696 = NDP = 38,5 RDP = 43,439
AF = 62.605.504 = NDP = 39,5 RDP = 42,278
AF = 67.421.312 = NDP = 40,5 RDP = 41,234
AF = 72.237.120 = NDP = 41,5 RDP = 40,241
AF = 77.052.928 = NDP = 42,5 RDP = 39,294
AF = 81.868.736 = NDP = 43,5 RDP = 38,390
AF = 86.684.544 = NDP = 44,5 RDP = 37,528
AF = 91.500.352 = NDP = 45,5 RDP = 36,703
AF = 96.316.160 = NDP = 46,5 RDP = 35,913
AF = 101.131.968 = NDP = 47,5 RDP = 35,158
AF = 105.947.776 = NDP = 48,5 RDP = 34,432
AF = 110.763.584 = NDP = 49,5 RDP = 33,737
AF = 115.579.392 = NDP = 50,5 RDP = 33,069
AF = 120.395.200 = NDP = 51,5 RDP = 32,427
AF = 125.211.008 = NDP = 52,5 RDP = 31,809
AF = 130.026.816 = NDP = 53,5 RDP = 31,214
AF = 134.842.624 = NDP = 54,5 RDP = 30,642
AF = 139.658.432 = NDP = 55,5 RDP = 30,090
AF = 144.474.240 = NDP = 56,5 RDP = 29,557
AF = 149.290.048 = NDP = 57,5 RDP = 29,043
AF = 154.105.856 = NDP = 58,5 RDP = 28,547
AF = 158.912.664 = NDP = 59,5 RDP = 28,067
AF = 163.737.472 = NDP = 60,5 RDP = 27,603
AF = 168.553.280 = NDP = 61,5 RDP = 27,154
AF = 173.369.081 = NDP = 62,5 RDP = 26,720
AF = 178.184.896 = NDP = 63,5 RDP = 26,299
AF = 183.000.704 = NDP = 64,5 RDP = 25,821
AF = 187.816.312 = NDP = 65,5 RDP = 25,496
AF = 192.632.320 = NDP = 66,5 RDP = 25,113
AF = 197.448.128 = NDP = 67,5 RDP = 24.741
AF = 201.263.836 = NDP = 68,5 RDP = 24,379
AF = 207.079.744 = NDP = 69,5 RDP = 24,029
AF = 211.895.552 = NDP = 70,5 RDP = 23,688
AF = 216.771.360 = NDP = 71,5 RDP = 23,356

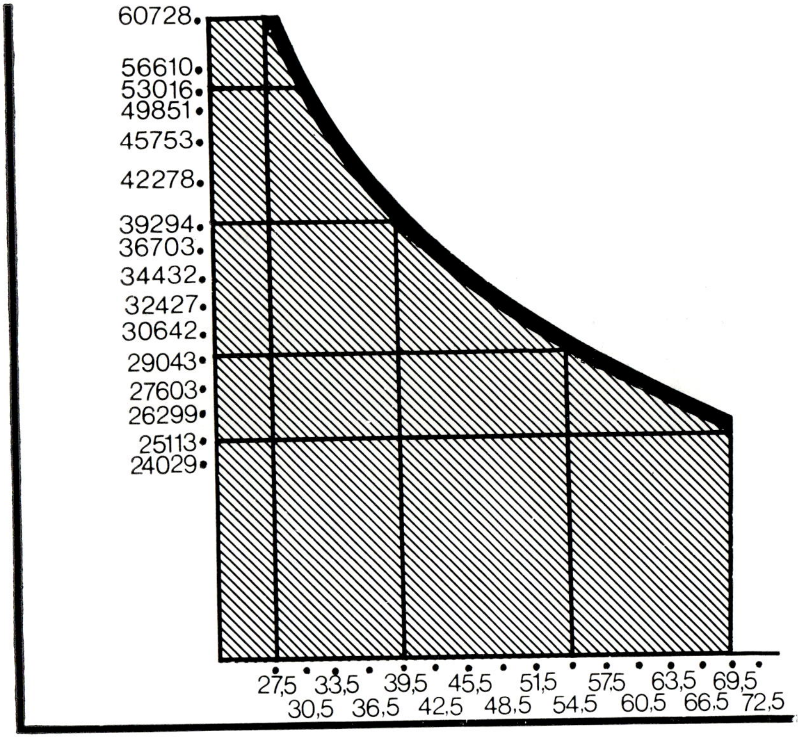

S = 1,670.000 = NDP = 26,5 RDP = 63,010
S = 1.676.253 = NDP = 27,5 RDP = 60,954
S = 1.682.506 = NDP = 28,5 RDP = 59,035
S = 1.688.753 = NDP = 29,5 RDP = 57,245
S = 1.695.012 = NDP = 30,5 RDP = 55,574
S = 1.701.265 = NDP = 31,5 RDP = 54,008
S = 1.707.518 = NDP = 32,5 RDP = 52,339
S = 1.713.771 = NDU = 33,5 RDP = 51,157
S = 1.720.024 = NDP = 34,5 RDP = 49,855
S = 1.726.277 = NDP = 35,5 RDP = 48,627
S = 1.732.330 = NDP = 36,5 RDP = 47,466
S = 1.738.783 = NDP = 37,5 RDP = 46,367
S = 1.745.036 = NDP = 38,5 RDP = 45,325
S = 1.751.289 = NDP = 39,5 RDP = 44,336

S	=	1.757.542	=	NDP	=	40,5	RDP	=	43,396
S	=	1.763.795	=	NDP	=	41,5	RDP	=	42,501
S	=	1.770.048	=	NDP	=	42,5	RDP	=	41,648
S	=	1.776.301	=	NDP	=	43,5	RDP	=	40,834
S	=	7.782.554	=	NDP	=	44,5	RDP	=	40,057
S	=	1.778.807	=	NDP	=	45,5	RDP	=	39,314
S	=	1.795.060	=	NDP	=	46,5	RDP	=	38,603
S	=	1.801.313	=	NDP	=	47,5	RDP	=	37,922
S	=	1.807.566	=	NDP	=	48,5	RDP	=	37,269
S	=	1.815.819	=	NDP	=	49,5	RDP	=	36,683
S	=	1.820.072	=	NDP	=	50,5	RDP	=	36,041
S	=	1.826.325	=	NDP	=	51,5	RDP	=	35,462
S	=	1.832.578	=	NDP	=	52,5	RDP	=	34,906
S	=	1.838.831	=	NDP	=	53,5	RDP	=	34,370
S	=	1.845.084	=	NDP	=	54,5	RDP	=	33,854
S	=	1.851.337	=	NDP	=	55,5	RDP	=	33,357
S	=	1.857.590	=	NDP	=	56,6	RDP	=	32,877
S	=	1.863.843	=	NDP	=	57,5	RDP	=	32,414
S	=	1.870.096	=	NDP	=	58,5	RDP	=	31,967
S	=	1.876.349	=	NDP	=	59,5	RDP	=	31,535
S	=	1.882.602	=	NDP	=	60,5	RDP	=	31,117
S	=	1.888.855	=	NDP	=	61,5	RDP	=	30,713
S	=	1.895.108	=	NDP	=	62,5	RDP	=	30,321
S	=	1.901.361	=	NDP	=	63,5	RDP	=	29,942
S	=	1.907.614	=	NDP	=	64,5	RDP	=	29,575
S	=	1.913.867	=	NDP	=	65,5	RDP	=	29,219
S	=	1.920.120	=	NDP	=	66,5	RDP	=	28,873
S	=	1.926.373	=	NDP	=	67,5	RDP	=	28,538
S	=	1.932.626	=	NDP	=	68,5	RDP	=	28,213
S	=	1.938.879	=	NDP	=	69,5	RDP	=	27,897
S	=	1.945.132	=	NDP	=	70,5	RDP	=	27,590
S	=	1.951.385	=	NDP	=	71,5	RDP	=	27,292

From these two series the preceding and following diagrams have been obtained.

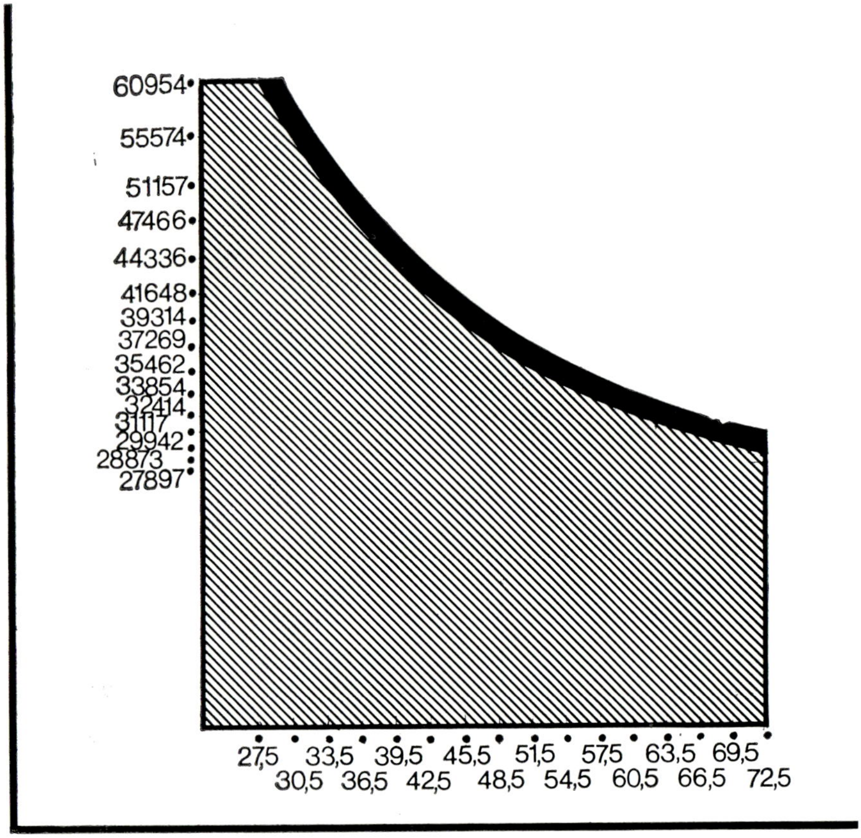

We have expressed the « correlation between indirect cost and revenue of the point of distribution » in terms of the following: a constructed model, a series of algebraic expressions, a mathematical calculation performed on the basis of solid data and a graphical representation of a function with two curves. We consider it necessary to give a representation following the methods systems' dynamics.

DDC	Direct distribution cost
DPC	Direct costs per point of distribution
P	Profits decided upon
S	Sales in N° of product units sold
y	N° of units needed to be sold to cover all non distribution costs
x	N° of units needed to cover DDC
NDP	N° of distribution points needed to get P
RDP	Average return of each distribution point
RU	Average return of each unit sold

The symbol P (management profit) has been included for the completeness of the representation, but it is not relevant in the correlation. Its role in the formulas is superflous from this point of view. We have introduced the symbol P to underscore also with the system dynamics block diagram that profit plays a role of practical and indicative importance even in a planned system no matter what the purpose and name given to it. In the ultimate analysis, P is nothing but a cost.

YIELDING CAPACITY OF A PUBLIC TRANSPORT VEHICLE CALCULATED IN KM/PERSON

The problem to which the following schemes and formulae refer is as connected to the settling of the population with its distribution between cities and countryside and division in industrial and rural activities, as to the problem of the distribution network.

We shall consider here only the influence of planning costs on as important a service as public transport, which represents the only viable alternative to private means of transportation in the future because of the increasingly dangerous ecological consequences sparked off in this area and it is indispensable to keep planning costs down in order to provide marginal services.

Scheme I — Finding the theoretical number of the vehicles necessary to achieve the prefixed result, all other management data being known.

We indicate with IC the indirect costs required to manage a public transport as a whole.

We indicate with c the average contribution.

We indicate with CV the direct cost for each vehicle per Km/person conveyed.

We indicate with U the management profit.

We indicate with S the total service rendered expressed in Km-person.

We have:

$$\frac{IC + U}{c} = y$$ = number of Km/person needed to cover the costs not tied to the vehicle or not included in the calculation of the average contribution per Km/person.

$$\frac{CV}{c} = x$$ = number of Km/person needed to cover direct expenses of the vehicle.

$$\frac{S - y}{x} = NV$$ = number of vehicles compatible with the appointed result of the management.

$$\frac{S}{NV} = rv$$ = average revenue from the vehicle expressed in Km/person.

Scheme II — Finding the correlation between the fall in indirect costs and theoritical number of vehicles.

Let AF be the amount by which indirect costs fall:

$$\frac{IC - AF}{c} = y$$ = number of Km/persons needed to cover all costs independent of the vehicle or not included in the calculation of the average contribution per Km/person.

$$\frac{CV}{c} = x$$ = number of Km/persons needed to cover direct expenses of the vehicle.

$$\frac{S - y}{x}$$ = number of the vehicles compatible with the appointed result of the management, after the fall in indirect cost resulting from management savings.

$$\frac{S}{Nv} = rv$$ = average revenue per vehicle after the expected fall in indirect costs.

Scheme III — A précis of the correlation between AF, NDP, RDP, supposing $AF\ 1 < AF\ 2 < AF\ 3 < AF\ 4$ and holding S constant.

S AF 1 NDP 1 RDP 1
S AF 4 NDP 4 RDP 4

The Nv'o will take on values:

$$NDP\ 1 < DNP\ 2 < NDP\ 3 < NDP\ 4$$
$$RDP\ 1 > RDP\ 2 > RDP\ 3 > RDP\ 4$$

Scheme IV — A précis of the correlation between S, NDP, RDP supposing $V\ 1 > V\ 2 > V\ 3 > V\ 4$ and holding the AF's constant.

S 1 AF Nv 1 rv 1
S 4 AF Nv 4 rv 4

The Nv's will take on values:

$$NDP\ 1 > NDP\ 2 > NDP\ 3 > NDP\ 4$$
$$RDP\ 1 < RDP\ 2 < RDP\ 3 < RDP\ 4$$

The schemes and the correlation between some of the essential data relevant to the activity and services planned and carried out in the programmed system could, of course, be greatly elaborated. We cannot however do it here. To follow up the question would mean to stress the necessity of holding the costs within a model which includes the economicity of the results and the priorities in the needs of the population. Furthermore, all this should be carried out within the framework of a correct application of what we have called the « comparative usefulness parameter ».

It is certainly impossible, even through the most careful of investigations and the adoption of the most modern techniques, to obtain a decision-pattern which may yield optimal solutions. One can only hope for good enough approximation.

The interplay of forces in the free market is unpredictable, but the presence of some pointers are ensured by the automatic nature of this system. In programmings these pointers are obtained through the reduction of a vast amount of data, which imply several estimates that can only evolve in the reality of human speculation.

EXAMPLES OF ALTERNATIVE CHOICES AVAILABLE BETWEEN TWO DISTRIBUTION SYSTEMS

Comparative waste: returnable and non-returnable bottles

The table that follows provides statistical data on consumption of wine, mineral water and soft drinks for the city and province of Venice, expressed in litres and followed by the number of bottles (of all sizes which are used to handle the liquids). We see from the table that the total quantity of all drinks considered is 161.5 million liters. The number of bottles in which the drinks are traded is 181 million. The stock (by stock we mean the number of bottles in circulation, which allow the annual consumption to take place for example it may be sufficient to have 20 empty bottles to support an annual cosumption of 100 bottles of wine) of bottles which the producers, the distributors, the shopkeepers, and eventually the consumers have at their disposal is 34 million. The number of bottles for wine, mineral water and soft drinks presentlys quandered is about 9.75 million; by squandered we mean broken and disposed of, whether shattered or not, in the garbage, on the beaches, on meadows, in rivers, in lagoon, in sewers.

The table intends to underline the fact that if — within given-bounds all drinks considered were consumed in non-returnable bottles, the number of squandered bottles would be 181 million rather than the present 9.75 million; this would intensify enormously the gravity of the situation with regard to the collection of garbage, the choking of sewers. Even if only half the trade were performed with no return bottles, the squandering would be 90.500 bottles — representing still an enormous ecological upset.

If all trade of the three drinks considered were performed using no return bottles, taking the average weight of a bottle to be between 300 and 350 grams, the weight of the total waste would be about 55.000 tons. Since the trucks which can circulate in the city can be loaded with approximately 4 tons, and since we are dealing with low density and not closely packed goods, we must assume each truck can take 2 tons, so that 22.500 trucks would be needed to dispose of the waste.

It is clear that what has been mentioned must be taken into account in choosing between the two systems, and traditional criteria are here replaced by considerations which include more far-reaching goals.

	Litres consumed	By means of number of bottles	Stock empty bottles	Squandered bottles or broken
WINE				
City	35.000.000	25.000.000	5.000.000	1.250.000
Province	75.000.000	50.000.000	10.000.000	2.500.000
MINERAL WATER				
City	8.500.000	12.000.000	3.000.000	1.000.000
Province	24.000.000	34.000.000	5.500.000	2.000.000
SOFT DRINKS				
City	5.000.000	15.000.000	2.500.000	800.000
Province	14.000.000	45.000.000	8.000.000	2.200.000
	161.500.000	181.000.000	34.000.000	9.750.000

CHAPTER FIVE

OPPOSED VIEWPOINTS

OBSERVATIONS ON FORECASTING FEASIBILITIES

When dealing with programming and, in particular, environmental programming, only rarely is the accuracy with which such programmes can be realized mentioned. The planning technique can be considered to have originated in military strategy and science of the eighteenth century, in which it was applied to win battles. This technique was later used in other areas and it reached full codification in the industrial sector. It was also applied, with an inferior degree of success, to economics, and, with no success at all, to city-planning. There exists even an economic doctrine based on this method called « the ungovernability of large systems » and examined by Francesco Forte.

Talking about planning the environment, is equivalent to talking about planning the world environment. An extension of the method on a simple scale (expanding with the size of the phenomenon) does not seem legitimate.

To formulate a plan the principles must first be fixed, defining the goals toward which the phenomenon should evolve, as well as the means to be employed in their pursuit. Then, all parameters which regulate the phenomenon must be studied and correlated to the proposed principles, and consequently action may be taken.

The principles chosen for the construction of a logical conceptual structure are considerations which must be accepted, because of their very nature, as acts of faith, and also because it is possible to construct a logical structure with their opposites. Examples of this are plentiful, from wave mechanics to quantum physics, from heliocentrism to the orbiting of constellations, from democracy to absolute monarchy. To say that one is better than the other is to indulge in an act of faith, even though in some cases such an act is skilfully masked by experimental proofs.

In general only social principles are regognized as acts of

faith, where the concept of « justice » is, among other things, included as something which really exists.

A more accurate approach is to consider the principles as working tools and to refrain from the absurdity of comparing them, which is to say that no moral or ethical value should be attached to them. This assertion implies the impossibility of making logical choices and it is necessary therefore to rely on visceral choices. For example, it has been said that if every Italian were to possess 5.000 meters of land to build a house on, all the plains would be filled with construction sites.

A negative bias is implied in this statement on fact. It would not be so, however, if it were framed within the quality-of-life principles of the Italian people, such as representation, the possibility of doing some gardening, the necessity of cultivating vegetables, the agricultural exploitation of mountain areas and so on. It would be quite possible to develop a social structure on such basis. To accept it or not would be a visceral, or habit-formed choice.

Another example is the right to property, which can only be assigned ontologically. In favour, there will be all those who have property or wish to have some, according to the rules of the right to property; against, there will be those who have no property at all and do not hope to gain any according to said rules.

Quite another problem is that of identifying the parameters and their interrelations. Our present state of knowledge permits us to work with only a limited number of parameters. When scientists were first confronted with a problem defined by a very large number of parameters, they found it impossible to apply Lapalician principles of causality.

Max Born said « The hypothesis of absolute observability seems to me to exist only as a postulate of the imagination which cannot be realized in the real world ». He then proceeds to look for the impossibility of satisfying the hypothesis in the quantum nature of energy and in the impossibility of making a measurement without disturbing the system « measured », deeming it possible, on the other hand, with due caution, to verify the hypothesis for macroscopic phenomena.

The phenomena we should consider are those which in space and time are characterized by dimensions of the order of those of man. Everything that is too far from normal human conditions can only be known through those aspects which can be mapped back to these dimensions; the farther one goes away from them, the more superficial our understanding and the more limited our casual knowledge. Phenomena extremely far from human conditions remain fundamentally unknowable. Some examples follow:

1) Casting a well-made die a very large number of times, it is found experimentally that each face appears about 1/6 of the time. This is to say that based on the phenomenon of a set of very many throws such a regularity can be established. No particular prediction can be made about an individual throw, however, since the outcome is determined by such trivial factors, the energies which enter are so small, that we cannot assess them and place them within a logical framework, so that whereas we were able to verify the phenomenon of a large number of throws, we can say nothing about any single throw.

2) An event whose lifetime is very long compared with the time of observation can be interpreted as one deems best, since it is not possible to repeat it (e.g. glaciations, stellar explosions, life of the sun).

3) An event whose lifetime is very short remains indeterminate, since it is not possible to construct a conceptual category based upon it (stationary state — e.g. motion of an electron), which is to say that it is not possible to place it in relation to phenomena which occur on a human time-scale.

4) Phenomena which take place over a wide space have an evolution which remains unknown (e.g. the life of the ocean).

One must, therefore, in developing a plan, assess the dimension of the phenomenon to be planned, and to consider the time for which the plan is to apply to the phenomenon. All other parameters which must be present in its logical structure depend from these two as well as from the casual or unknowable nature assumed by the phenomenon to be planned.

It can be deduced, therefore, that the phenomenon to be planned must be contained within well-defined boundaries if the result planned for a given future time corresponds to the evolution of the phenomenon during the period. Outside these limits — but not too far — the probability of a certain event can be discussed.

We consider it useful to complete this discussion with a few observations.

When talking about a principle, it is appropriate to recall the thought of Werner Heisenberg, one of the founding fathers of quantum mechanics, and expositor of the uncertainty principle. The concepts of classical physics exercise the same role as the a priori forms of intuition in the philosophy of Kant. In the same way that Kant calls aprioristic the concepts of space time or causality, the concepts of classical physics are an a priori basis for quantum experiments.

The same statements are made about the M.I.T. report by the Club of Rome. « Any model can only include a limited number of variables, therefore the interactions are only partially studied. Even a simple model of the world permits the study of the effect of a change in the fundamental hypothesis, or to stimulate different policies, to see how such changes influence the behaviour of the system in time. To conduct such experiments on the real world would be in many cases impossible, and in any eventuality every time consuming and costly ».

Concerning phenomena which manifest themselves in long periods of time Pierre Auger writes that unfortunately the verification of some theories would require conditions which cannot be realized, for example extremely precise observations for a very long period of time, such as a billion years.

Concerning phenomena which take place on a very short time scale, Heisenberg himself points out that the forms taken by our intuition do not adapt themselves to situations where velocities near that of light present themselves.

Concerning phenomena extended over vast regions of space, we quote Auger: astrophysics has shown that a part of the universe will always remain inaccessible no matter what means are employed to gain knowledge of it.

MAN AS A POSSIBLE OBSTACLE TO PROGRAMMING

The image we have of man as a psycho-physical entity determines the formulation of the programs which, in a planned economy, must tend to satisfy the needs implied by his nature. But the traditional concept of man may be considered out of date, and in any case not conforming to his genetic reality.

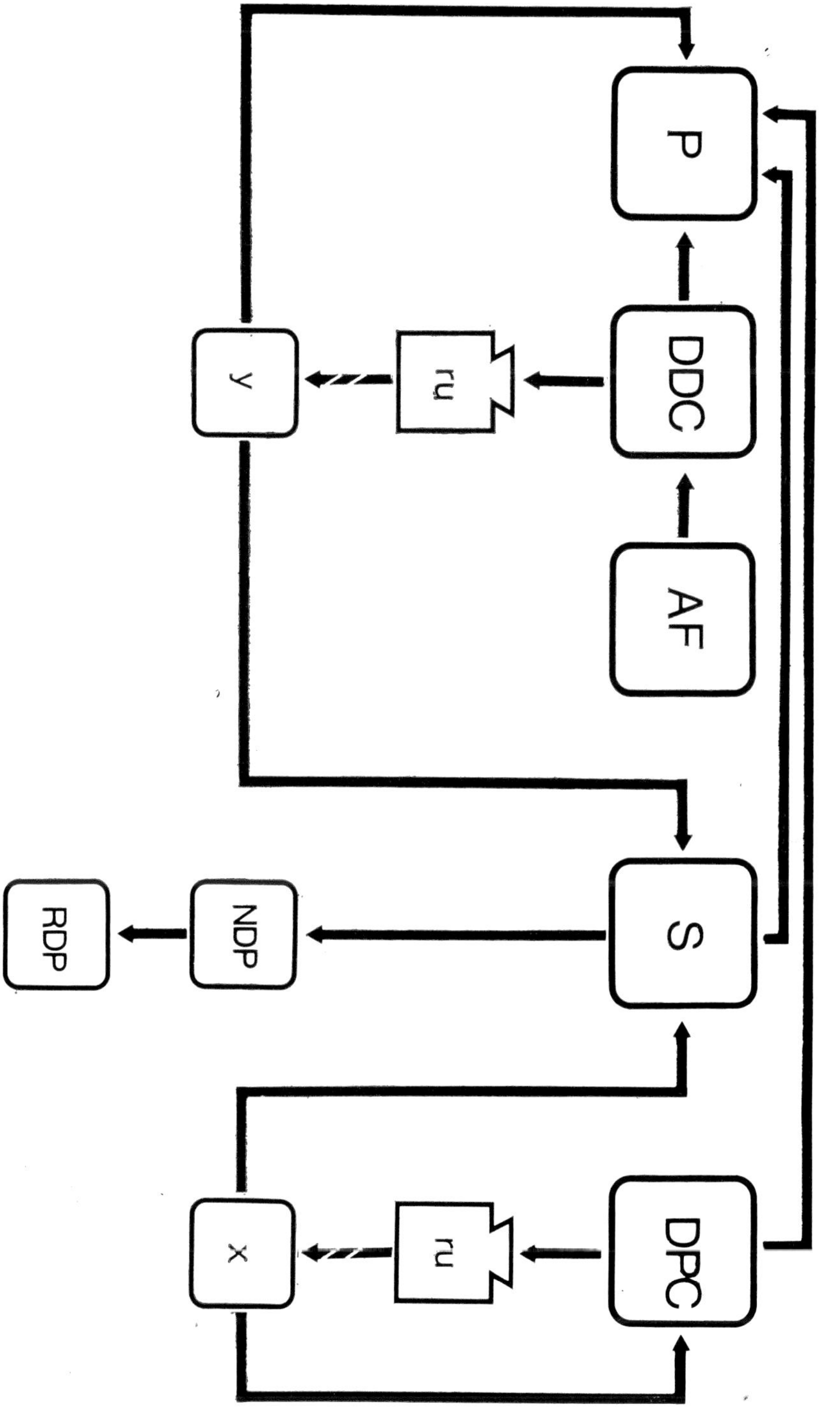
P
DDC
AF
ru
y
S
NDP
RDP
ru
x
DPC

In the traditional view man is the center of the universe, he is the privileged creature made in the image of God, he has a high position in the universe, and is on good terms with the Almighty. Catholics or Protestants, Jews or Mohammedans, believers or atheists, men over all have faith in the prestigious image of themselves. Such a high ranking position in the biological world brings with it the window-dressing of noble ideals, in a context of self-esteem which justifies rights and priorities over the rest of the animal world. In the whole of history the relationship between men and animals is based on the assumption that we, and only we, are intelligent beings; man defines himself as the only rational animal. We have therefore claimed many privileges, and we have created two different ethical systems, one applicable on a man-to-man basis, the other valid for a man-to-beast situation.

Bill Gilbert in « The language of animals » expresses himself as follows. The new way of considering animals comes from the most recent study on free animals and not on the unhappy specimens kept in captivity. Understanding their behaviour and their real motivations is the most important discovery made by man in the last 40 years. It is much more important than the discovery of atomic energy and many other breakthroughs.

The reason is easily explainable: the study of animal behaviour, theis rituals, their life styles, is the only method conducive to an understanding of man: to understand him in this true reality, as carrier of an extremely vast biological inheritance.

To know animals and their behaviour means tearing away the veil which hides the true nature of man, leaving the concept of central observation to look at ourselves as just one of the many participate in the ancestral network that comes to us, not from a precise source which we identify with primates, but from a heritage 400-million-years-old.

But after all, who is man?

We may find an answer in a precis of Robert Ardrey's thoughts. Man is a hunter for whom killings is a natural instinct. The liaison between a developed brain and carnivorous habits made man into a genetic possibility. The primate from whom we descend has instincts which demanded territorial acquisitions and protection thereof, a permanently hostile attitude toward neighbours; the formation of territorial groups as the principal survival cause for an essentially vulnerable creature. When the needs of a predatory existence meet the fundamental instincts of the primate, these intensify, conflicts become lethal, and territorial disputes become small-scale wars. The social group intended as a hunting and defense unit becomes more rigid in its friendships and enmities.

The quest for man may lead us even further inside the entrails of a quest for history through layers of experience preceding primates, preceding even mammals: we will find the instinct to dominate one's fellows to defend what is regarded as one's own, to eat and avoid being eaten. Even in the more remote levels of human nature we find traces of animal nostalgia, of fear, domination and rank. This is our heritage.

We have an innate quality of self-awareness not to be found in other animals: whether this is a consequence of a developed brain, or a characteristic inherited through our ancestors, remains unknown. Consciousness has been the cornerstone in human behaviour for the development of civilization.

But consciousness is a two-sided coin and one of these faces is dark. The polarity: friendship-hostility, is a property of the primates, who could not survive without enemies.

Consciousness organizes both hate and love. It can induce the roman martyr to die for human fraternity, but at the same time it can induce Christian armies to march and slaughter their fellows. Consciousness is, by its own nature, territorial and totally amoral. It is possible to fool ourselves into thinking that it orders actions in the interests of human good. This may be the case; yet with equal force it enjoins men to act in the interests of human evil if this evil serves the interests of the society to which one belongs. God is never so much in fashion as in wartime.

By its very nature, consciousness cannot tackle universal problems. Equivocal in its necessities, monopolistic in its territorial takeovers, it denies any relationship between antagonists.

To expose such perhaps extreme thoughts, but not without foundations and nowadays widely shared by many scientists, was necessary, above all, for the conclusion relevant to our treatment. Let us look at some of these conclusions.

Studies performed at Harvard University have revealed that delinquent youths, compared with youngsters of analogous social extraction, have superior energy and physiques and among delinquents there is a much smaller neurosis incidence. This, we observe, is in line with Freudian theories. If natural man, if the young hooligan does not have what society defines as conscience, it is because his instincts have known few of the inhibitions generated by society. Society prides itself in believing it has rejected the delinquent, but it is the delinquent who has refuted society.

So, society has but one problem: to condition a baby as soon as it is born. Each society, based on collectivism or individualism, has its techniques to harness its new wild citizens: fright and satisfaction, punishments and reward, dissapproval and praise, all to develop those instincts which adapt themselves to the ends of society.

But the last interesting point we would like to make concerns a singular position regarding the destruction of our species through the atomic bomb. It is evident that we can extend these considerations to the danger of destruction through pollution, the poisoning of the atmosphere, of food, etc.

Ardrey in drawing his conclusion states that if of the whole world population only 100 million survived, at least half would undergo mutations which would lead this branch to extinction, but at least half a million people would have characteristics superior to those of their forefathers. Ardrey's work is not a collection of eccentric ideas, it rests on the knowledge of now widely accepted facts, even if the conclusions are clearly and completely his own.

It suffices for our purpose, to observe that this is not an isolated voice, but that the image of man that we have so rapidly sketched, is not far from the convinctions of other scientists. There is, in any event, no doubt that the traditional image is much less trustworthy.

In the study of problems which concern man, even only from an economic point of view, a different, but probably more authentic view of human nature must be taken.

Accepting the extreme point of view championed by Ardrey, the solution of the problem could be to let the human species develop spontaneously. Evidently we beg to differ.

Chapter sixth

FOR AN IMPROVEMENT IN THE HUMAN CONDITION

One of the most serious and most rarely mentioned forms of pollution is that suffered by intelligence and will-power. Let us go back for a moment to the mechanics of Herbart's ideology, for whom ideas can be classified according to their intrinsic quality and the degree of their intensity. Ideas are also tightly connected with time.

Herbart constructs a mechanics of ideas-representation (Vorstellungen). Two ideas of opposed quality inhibit each other according to their relative strenght. The prevailing of one idea over the other depends on its strength. The metaphysical reason for this lies in the unity of the soul, of which ideas represent its self-preservation. The strength of the idea is not to be confused with its quality. The ability in exposing an idea, its dramatic representation, repetition of the same idea, the most suggestive form in which it is represented, all condition its potential for survival in relation to the opposite idea.

It is sufficient, now, to mention the great influence of an idea which appears in the press, on the radio and above all on television, to understand how, whether it wants or not, the human spirit, to preserve its unity, must expell its polar opposite to that which is dished out for the longest time and with the greatest ability. This holds at least for the vast majority of cases.

The damage produced by the exponential growth of industrial production, which is, after all, damage produced by maximization in the capitalistic system and by the logic of profit, appears not only in the pollution of the biosphere, in the impoverishment of the resources to which necessarily turns for his survival, but it infests the ethological sphere, which is to say all of human behaviour, and first of all in polluting culture in the broadest sense of the word.

The profit logic leads to the creation of multinational business complexes of colossal proportions. Therefore, such a system, dominated by few centers of economical power (oligopolies) exasperates the instincts of each man driving him to satisfy an ever-increasing number of needs.

Consumerism, as a product of deteriorated cultures is imposed by an obsessive diffusion of behavioural stereotypes, maintained at every level by the information media. The school system, the press, radio, television, cinema, the publishing trade are completely at the service of a culture which sinks its roots deep into the capitalist system. The purpose is that of creating an individual who stands out as little as possible from the mass, whose needs are easily predictable and sufficiently uniform, according to the canons of mass production: an individual to whom new needs and consumption goods can continuously be suggested.

At this point it cannot be denied that some socialist regimes do nothing but retrace the methods befitting the capitalist world.

If we want to salvage the type of human civilization which has formed in tens of thousands of years, the point of departure can only be that set of acquisitions and ideological conquests which constitute the fundamental points of progressive conceptions, from the Enlightenment to the French revolution. It is evident therefore, that what we cursorily and approximately call socialist thought is the only conception which can be used as a basis for a future society, which will have to develop its own themes in the field of ethology to arrive at a control of consumption and at that social life-style which are the prerequisites for the survival of man.

It is, of course, impossible to foresee which systems, which new habits, which culture will actually be generated in the near future. We simply wish to indicate the point of departure, which can only be a complete awareness of the ecologic situation. This imposes a revision of the intrinsic tendencies in the societies of today both in the East and the West.

A prerequisite for any concrete action is the formation of a well informed and aware public opinion, kept up to dare on the great problems which confront mankind.

We live in a part of the world which is within the sphere of interests of the powers and conceptions of capitalism, and we can competently say that the informations delivered to the public are always tinged with parochial undertones.

The profit logic finds its limits in the expansion of its products and in the inevitable extension of its results. The linking and transmission of information among individual, on a private level will become increasingly easier; private cable television stations will go on the air ever more economically. Even if books are the prerogative of a few pictures are welcomed and accepted with extreme ease by all. It will be possible to divulge scientific breakthroughs more easily using pictures and words, especially by presenting them in more accessible forms.

The simplicity of the language, together with the concept of informing in a unitary, rather than frangmental fashion, the ease of diffusing information through living images and words will be the most important, if not the only weapon to transform society: a society without deceptions or utopias, without far-fetched promises, but with the full knowledge of what mankind can achieve at a particular historical moment, with the given technological developments and without prejudice to future civil progress.

The human condition, with its aspirations and unsurmountable boundaries, is the basis for a programming which may have some hope of success.

The problems inherent in the present situation, which express themselves in the most varied turbulence throughout the world, and therefore also in our country, have a common basis in the contradictions implied by the development of economies in the more advanced countries and the state of backwardness and insufficient level of satisfaction of the most basic needs found in other countries. The system of development, common to advanced countries, be it on the capitalist or the socialist footpath, affects the difficulties already mentioned, and foreshadows a collapse. At the same time the global needs of the population of the world, even considering food alone, are far in excess of what is available, so that the menace of hunger looms over mankind.

But a situation of this kind, in which the « richest countries tend to become always richer, and the poorest countries tend to become always poorer » is complicated by the damage which the lifestyle introduced in the more evolved nations imposes on the entire ecosystem, involving also the more retrograde populations. It is evident, therefore, that whereas part of the population which benefits from so called well-being is limited to the more advanced countries, the damage following upon this well-being is felt by all inhabitants of the world.

This is the general context; the facts that must be dealt with transcend the possibilities of an investigation which leads to the discovery of concrete remedies. It is already a lot if the research yields an overview of the tragic reality in this field, to be opposed to the optimistic attitude that had spread over America and Europe after the Second World War. It can be considered a success if today, in contrast to what was the case fifteen-twenty years ago, the dangers and threats that will menace us in future can be clearly perceived.

Never, in such a brief time, have we gone from a confident conception of so-called progress, of the possibility of improving the human condition in all its aspects, to the abandonment of many illusions and of a near-mystic faith in the possibilities of man, who seemed already on the road to conquering the universe. Today, rather than exalting the conquests, we make inventories of the damage and the price paid for so many disorderly whims.

Such a general introduction is important because only after grasping the principal points is it possible to arrive at an investigation of a particular fact. Vico expresses himself thus: « ... facts and affairs are more easily percieved in their reality by him who presents himself things through general and simple ideas, rather than by him whose mind is flooded with detailed forms ».

The situation of « well-being and development » under discussion involves the problem in its socio-economic context as a prerequisite to the transformations the general economic activities must undergo, to arrive at alternative systems in our field of interest, in the full light of the over all problem. We have attempted to do this by pausing over a new interpretation of costs and comparing market economy with programmed economy, seeking means of intervention and transitional systems, because only a gradual intervention can guarantee changes without damages, in a situation so delicate by the ecological factor.

We consider it useful to recall, as we did for Smith, Ricardo, Marx and many others, the thoughts of a great Italian thinker the superficial knowledge of whom, or rather the misinterpretation of whom, has prevented a justified revaluation. We wish to place him among the greatest revolutionaries. We cannot, in fact, not call revolutionaries those who rose above the conception consolidated in their lifetime, in order to express thoughts which are ahead of their times. Only a subtle investigation of the true nature of the facts of their period could lead them to a prediction of where contradictions existed in the world in which they lived.

We are, of course, talking about Nicolò Machiavelli. It must be remembered that he lived in a city and during a period that represent one of the moments of greatest prosperity in human history, be it economically, civically or culturally. In such a fertile period, Machiavelli was able to see the symptoms of dissolution, the seeds of decay. Machiavelli adopted a historical revolutionary approach: he starts from Polibio's concept of cycle through which monarchy degenerates into tyranny, tyranny becomes oligarchy, oligarchy passes into democracy, which corrupted by demagogy can only lead back to government by one man.

As Polibio attributes the passing from one political system to another to corruption and decadence, so Machiavelli sees in the latent corruption of Florence, even at the peak of her civil splendour, the presence of danger. He writes that Italian princes were over-concerned with their courts and their splendours and behaved with miserliness and arrogance toward their subjects, conceding anything only by « grace ». He understands that the people are split into two classes and opposes the leading class to the rest of the people, indicating thus the existence of two cultures; one for the governing and one for the governed. Considering the history of the Roman republic, he talks about a « mixed government », which must include the three forms of healthy government: by the Prince, by the wrangler and by the people, judging it more reliable for their mutual control.

If we think of the historical period in which this thought sprung forth, we cannot be surprised by how far Machiavelli's intuition took him, even with the difficulty of understanding his words, which refer to facts, institutions and circumstances so far from us.

The idea of bringing agricultural and urban culture together is one of the most modern solutions that can be imagined. The return of the urban population to agricultural activities is an indispensable condition for the protection of the environment, showing the degradation of areas now abandoned, and the recovery of agricultural activities which must be supported by a framework which civilizes life in the countryside. At the same time it is necessary to promote in parallel, complementary or related industrial activities, so as to create economically integrated zones, which guarantee a more rational ordering on both the territory and the population.

From what has been said it is clear that we see in the thought of classical economists the best starting point for the development of a new system which we have referred to as « planning system ». This is particularly apt since more than in politics we are interested in those structures which economics will presumably adopt to give a concrete shape to the new socio-economic reality.

We have often referred to the USSR and to the Russian five-year plans. We consider it useful, for the purposes of a more precise rendering of the present socio-economic situation, to make a comment on China, which finds herself in a very particular situation. China is governed by a socialist regime, and is today very backward, judging her by the yardstick of development, with respect to Russia, having in common with the latter a similar basic ideology.

This backwardness can, however, be an advantage, since it gives the Chinese government a greater flexibility than that enjoyed by the Russian Poliburo.

It should be pointed out that Russia nowadays imports wheat and that agriculture is one of the sectors in which the USSR has not been able to fulfill its expectations. The justifications and reasons may be several, but the problem for Russia remains. China, also on the basis of the Russian example, has understood the problem and bears it in mind.

Agriculture, both because it satisfies the major part of the food requirements of mankind, and because the population settled in the country is the best means of defense of the land, remains the crucial problem for the destiny of present-day civilization.

EVOLUTION OF THE CONCEPTS OF SCIENCE AND NATURE

Nature has been variously conceived of in the course of time, first as expressive of harmony of the universe (in a fundamentally religious attitude towards the relation between man and nature). Researchers then objectified nature, and already with Galileo and Newton a keen interest in details emerges, seeking the laws and the relations which regulate certain phenomena. There is an effort to reach the universality of the laws, and from a purely descriptive process, we pass to an attempt at explaining nature to investigate the ultimate essence beyond tangible appearances. Science is the attempt of man to penetrate and understand the laws of nature.

Until the turn of the century scientists and the public had an essentially critical attitude, a systematic doubt regarding experiment. Today, on the other hand, the attitude of the public, confronted with continual discoveries and achievements, which only a few decades previously seemed unattainable, has turned around one hundred and eighty degrees. The public has begun to believe that everything, sooner or later, would become possible and that science would have dominated and shaped nature to its will.

But another reason for this attitude is the extreme complexity of science. Today, even for an educated person, it is already significant to be able to understand the conclusions and concrete achievements scientists have arrived at. There is nobody, after all, who may aspire to a complete and detailed understanding of the knowable.

We have already said it, but, we wish to stress it here, that science is seen today as a series of myths, which people regard with satisfaction, amazement, anxiety, but without the least spirit of criticism, since they are in no position to understand it. After all, mankind has become detached, in some sense alienated from nature, and there exists a great misunderstanding of the concept of nature and of that which is natural. It is indeed difficult to feel part of nature and at the same time to dominate it.

There was a time when man felt nature to be a set of conditions to be overcome in order to survive. Nature was the object of his struggle and his research; nowadays man has gone beyond this antithesis. There was a time when some security could be derived from the faith in the greater laws of nature, nowadays scientific achievements are soon outdone and modified. There is nothing absolute, not even in the exact sciences, or at least there are only valid acquisitions in well-defined and limited fields.

It is therefore evident that we can no longer think of a science that provides an idea of nature which could constitute a foundation based on any other kind of certainty. Science no longer attempts to give a picture of the world, but is itself involved in the relationship between man and nature, which of course cannot be constant, but is in a state of perpetual becoming.

If all this is true, it is clear that man can only suffer from the extreme vulnerability of his knowledge. It should also be noted, in this respect, that the enormous development and diffusion of the means of communication, especially audiovisual ones, has brought some advantages, but also some disadvantages. In the final analysis, man has better ways of communicating with his fellows, but the fashion and the situation in which he does so increase the difficulty of mutual understanding.

It is evident that the problem of knowledge has reached a critical point: we dare say that here, also, we have a development or human progress crisis — not materially, but spiritually.

We have already seen what the present attitudes are toward the material-development crisis, and after all a certain consistency can be perceived in that direction. The crisis of man face-to-face with science and nature draws us inevitably into the realm of philosophy. It would of course be out of place to deal here with gnosiological or ethic problems; we would like however to present some considerations which touch our research concerning priorities.

The simplest and most useful question that can be asked for our purpose is this: what is it that man, being what he is, wishes most for himself? Clearly an enormous number of answers can be given, according to the philosophy, religion or spiritual attitude of the person questioned. We could suggest, according to the most modern thought, *a tendency towards self-realization, the preservation of internal order, and an actual dedication towards others.*

It would be out of place to express opinions or to review the principal answers that could be given; we have indicated a few by way of example. Instead we wish to stress that every man should be in a position to give his own answer. The choice can only be an individual one. Society must only guarantee the freedom of this choice. No State, no Church, no political system should therefore impose noble ideals or other outlooks on life.

We hold however that education and science can help man in his choice. This can be done also by tackling fields which have only now begun to be explored through valid criteria.

First there is Kinetics, the science which deals with models of behaviour and non-verbal communication. In this area one of the most recent breakthroughs relates to the relation which exists between animal and human forms of communication. To mention just one of the many questions dealt with by this science, we quote the sense of territory, which is common to men and animals, with the difference that where animal species have come to terms with this imperative, for man it remains one of the principal sources of conflict, not only on a macroscopic level but above all in the relations among individuals.

The science of human behaviour opens totally new insight into the ramifications of certain fundamental human needs which do not always reach the level of consciousness, but which lead, when suppressed, to aggressiveness and misunderstanding towards oneself and towards others.

The subject is fascinating, but we can only stop briefly on another point of great interest to the present work. According to the first notions of social psychology, human impulses were always taken to be innate. With the dawn of ethology, the science that deals with comparative behaviour, it was understood that for animals, too, the impulse to act was not dictated exclusively by primary physiological needs. A man does not behave in a given way exclusively on the basis of the genetic heritage passed on to him by the ethnic milieu from which he derives. The behaviour of individuals, even of different races, is above all determined by the cultural conditions in which he lives.

In other words, a Japanese, an Indian and a European, placed in a situation of squallor and misery, behave quite similarly, showing that the objective and immediate situation in which a man lives influences him much more than his ethnic heredity. This observation can be extended to many other fields, especially that of morality and decency. It implies therefore a very different approach to the problem of crime, maximizing the influence of the milieu and the satisfaction of basic needs, both material and psychological.

In the search for the fundamental needs and essential characteristics of man, a science which may validly explore unchartied territories is semiotics.

The study of languages and all forms of expression is of fundamental importance because it proposes a reconnaissance of human history from a point-of-view less liable to false analogies that may be observed among different peoples at different times can only bring us back to the essential motives in human experience. Myths, folklore, drawings, all hold messages which only nowadays are beginning to be interpreted intellectually. The evolution of the sign and language is the evolution of thought. Sign and languages are nothing but a laboured end-product of that which preceded the organization of the mental processes in man.

So, it is necessary to relive these ancestral experiences and go back to the roots of behaviour and to the first expressions of thought to find a more authentic feeling of what man is, so that he may know himself beyond the superstrata which progress has imposed on him.

Lastly we wish to recall that parapsychological research is a very important innovative field, aimed at learning about unknown faculties which are present in animals and in some exceptional individuals, which man has lost, but which he could regain.

In any case, to tackle problems and questions which might seem marginal is always positive when these lead to introspection and to man's discovery of his place and his right dimensions.

There has always been an attempt to reconcile science with morals; nowadays it is necessary to reconcile morals with authentic, scientifically-acquired knowledge.

If man will manage to understand that some values have been imposed on him and if he will then suceed in pushing beyond them, he will then also have the moral freedom to adapt to concrete situations, without discarding his constant tendency to overcome difficulties by means of that supreme biological virtue of his of adaptation to environmental conditions.

Balzac had one of his characters very cynically say: « There exist no principles but circumstances ». For man it is, after all, really so: to be himself, to continue to survive, to follow the supreme imperative of the species, it is necessary for him to adapt; to do so he must above all be truly himself in the fullness of his genetic heritage.

ENTROPY AND BOUNDARY CONDITIONS

The concept of entropy can have very wide applications. Since we use it in a socio-economic mode of research, the definitions we will introduce have only that purpose of clarifying why this law of physics proves useful in understanding the premises from which we started.

The law of increasing entropy states that in converting one type of energy into another, some must necessarily be lost. For example, seeking to convert electrical into kinetic energy, it proves impossible to do so completely, since a part of the initial electrical energy — rather than becoming kinetic energy — is dissipated as heat.

We repeat that the laws governing entropy are valid throughout nature and the concepts involved have the widest potential applications.

We must now keep in mind that energy exists in nature in two forms: useful (or free) and unusable (or bound). Man can only use the first kind.

The energy contained in a piece of wood is free, and man can use it by extracting heat. The energy resulting from the difference in temperature between the water in a container and its surroundings is also free.

There are so-called orderly forms of free energy. Bound energy is dissipated in a disorderly fashion.

This kind of reasoning reflects an entirely human approach in interpreting the facts of nature. What we have said and what follows is nothing more than the representation created by our minds of facts which could be viewed differently by other than human minds.

It is futile to assume, in our study, of phenomena, that our logic, physical and chemical laws and so on, with which we classify phenomena, really exist. Logic, physics, chemistry, mathematics, geometry, are nothing more than the way in which our minds apprehend reality.

Ours is a subjective reality, as seen by the human mind.

Having said this, we return to the implications of entropy in the economical process. On the physical level the economic process does not produce matter-energy, and does not use it up, but captures it in order to return it ad infinitum into the environment. This is a totally physical interpretation of economics.

Stressing economics itself, we can talk about using up valuable resources and rejection of matter devoid of value. This could be stated differently: matter-energy is fed into the « black box » economics in a state of high entropy. This holds true, in our view, in that scrap is always of light entropy content. In fact all that has economic value has also an orderly structure, and therefore a low level of entropy.

Manufactured goods and fruit are of a low entropy, and they are both ordered structures.

It is our belief that the following association can be made. Low entropy-ordered structure quantitatively measurable in terms of economic value; high entropy-chaotic structure no assignable economic value.

It could be thought that living systems contradict the principles regulating entropy, since in fact in every organism exists the effort to maintain constant entropy. But in reality every living organism absorbs from its environment low entropy substances and thus inexorably increases the entropy of the system which includes the organism. This translates into a degradation of its structure, which expresses a situation rooted in the biological reality of life.

Man, cooking his foods, transforming natural forces such as the flow of rivers, in mechanical work, burning coal to warm himself, behaves differently from other living beings, but cannot release himself from the law of increasing entropy.

The most advanced technologies cannot help but increase the entropy of the environment, but the most important fact is that every process, be it biological or economic, is carried out at an expense which is greater than the result.

The increase in entropy is constituted by the pollution which the economic process creates in transforming valuable natural resources. From the point-of-view of economics we cannot reduce all this to a simple statement of physical fact. We must recognize that this flow of scrap generates some intangibles, which could generically be labelled the satisfactions associated with a certain standard of living.

The productive activities of man, and in general all economic activities, are based on the exploitation of the low entropy of the environment. The low entropy of the world which we live in is finite however, and when some low-entropy (therefore valuable) material is used up, it cannot be used a second time. The exhaustion of low-entropy resources signals, without possibilities for extention, the limits of survival of the human species.

Today it is understood that human life is rendered possible by the exploitation of limited resources. But even after having realized that human life is based on the exploitation of the low entropy found in nature, the illusion has continued that self-producing sources of energy can be found.

Consider now that free energy, which is to say low-entropy materials, has two origins, namely the mineral deposits existing of earth, and the flow of energy from the sun, which, through the phtosynthesis of chlorophyll, constitute the foundation of all life on earth. It has been a fundamental error to maximize the exploitation of environmental resources rather than using what derives directly from the sun's radiation.

In agriculture, animals which use as a source of energy chlorophyll photosynthesis, have been substituted by tractors, that consume mineral resources. The same can be said of fertilizers, artificial ones having been substituted for natural ones.

Between the two natural sources of low entropy, man has chosen to exploit the more readily exhaustible one, committing thus a serious error, and giving rise to a style of life which renders the error more serious.

In the course of organizing a research or a study, many different points can be touched upon. In this work we have attempted to determine the present situation; find the key environment factors; give a correct direction to the problem; look for alternative solutions; consider the means of implementing them, and keep in mind the boundary conditions. In this part of the work, having paused to raise considerations which reach beyond the purely economic aspect of the problem considered, we wish to pass on to the way in which limiting factor manifests itself in connection with ecological problems.

Living beings, plants, fish. animals, microorganisms etc. can live only if one or more of the factors needed for their development becomes depleted in the environment in which they reproduce. In other words, among the reguired elements, the one which is least abundant must not fall below a certain level; otherwise that set of circumstances which renders life possible cannot be created.

In practice it is this condition which makes possible the existence of an organism in a given environment. Of course, this limiting factor is to be measured with a sufficient margin for error, which determines the limits within the environment may be considered favourable. For example, the temperature must not vary more than a few degrees either way for an environment to remain hospitable to a given kind of organism.

Extending these considerations would take us too far afield. It is evident, however, that for man, too, a limiting factor exists, even though due credit is given to his ingenuity and his great adaptability. Man's ecological adaptability is also greater than that of many animals because of the fact that he is able to transform environmental conditions and can hence survive in surroundings which originally were unfit to support human life.

This, of course, is for a limited time; at first man extends his habitat, then — as we have seen — he destroys the possibility of survival.

As far as our problem is concerned, the limiting factor can be variously identified, and even from a socio-economic point of view several possibilities can be envisaged.

In the context of the specific treatment of the problem we have developed, the limiting factor is the recalculated cost of products. If a situation is reached where our « integrated costs » become too high, clearly the limiting factor for a society such as ours extends beyond the acceptable.

We have only wished to mention this problem, which is in fact implicit in the treatment given, because it is interesting to see how laws are common and applicable to several aspects of reality. And this is perhaps merely the proof that what we accept as laws are but an identical fashion of classifying phenomena according to the form of our minds.

THE PROBLEM OF THE QUALITATIVE STANDARD OF LIVING

The way in which the necessities of life are realized, the methods adopted for the production of tools, and of the consumer goods needed by man, determine the social and cultural aspects, the esthetic and moral standards of human civilization at a particular moment. In a poor economy, society determines with great precision the priorities in its goals. A rigid scale of values results, in which the economic factor is seen as an immediate preoccupation, which determines and conditions all social relations and political solutions.

In societies such as those of the present day developed countries, in which basic needs are satisfied, one might logically suppose that economic objective might have lost much of their importance or at least that economic factors would be less determinant in the behaviour of human communities. Perhaps as a result of tradition and of the memory of the hard battle fought for generations for the basic necessities of life, this does not occur.

The prevalence of production confers great importance upon those who promote it or those who carry it out, so that economics is the most important factor in determining the social ladder.

The prevalence of the economic factor beyond the level which it might still hold nowadays, can be explained by the importance that is to be attached to the filling of a void which new values have not yet filled.

The capacity to produce and technological efficiency are still taken as a standard in comparing countries, and they constitute an evaluative criterion.

Since production and consumption are often dedicated to superfluous goods, it is hard to explain how a nation or a human aggregate can be valued on the basis of characteristics which have lost their primary importance.

It is useful also to note that attempts are made to maximize the production of nearly useless goods, when there is no doubt of the insufficiency of public service. In all developed countries, expenses for public services are looked upon with suspicion, and with the intention of limiting them, whereas nothing is done to slow down the consumption of consumer goods of secondary importance. Among other things, the inadequacy of public services sooner or later hampers the enjoyment of other kinds of goods.

In reality it is truly economic systems which govern men rather than being at their service; it would be better to say that it is the giant enterprises, the State apparatus, the private-power groups, which force men to live in a society modelled so as to allow the survival of these organisms. The great pressure exerted by the large organization is turned towards the entire society, creating a life style which in the ultimate analysis is one of iron discipline.

The strange fact is that men obey pressures which, in the most diverse forms and attractive guises the present system applies without cognizance that in reality these are impositions. The citizen buys with the illusion that he is fulfilling himself, that he is increasing his standard of living, and does not realize that he is responding to extremely banal and stupid invitations.

The basic economic necessities of long ago are transformed into a stimulus to consume with a revenge, and this is cleverly cultivated with the only purpose of holding together organizations and power groups which instead of serving the public enslave it for their own purposes. The real explanation of this can only be the lack of values which allow man to assert and realize himself on other than a purely material level. Above all there is the lack of a capacity to find ourselves placed in occupations of a general intellectual character, to which few are inclined these days.

If economic objectives continue to be the more important ones relative to any other, there will be no alternative to the present system, and no modification to the present social structure will be possible. These concepts, found originally in Galbraith, are shared by many other authors, and this ascertains the present relevance of the problem. In particular Alvin Toffler has often touched upon the subject, with unique effectiveness, and we shall often quote him.

The *Club of Rome,* in the appendix to the new M.I.T. book « Towards a global equilibrium », has treated the question, giving the appendix the title « The New Threshold ». The *Club of Rome* has preferred to stress the problem, rather than taking a precise position, since any such approach must of necessity have political overtones.

We wish, however, to make a consideration of fundamental importance before continuing to analyse qualitative standards of living.

In his economic activity, man has kept all those behavioural characteristics which emerged during his miological history, in the evolution from primate to the present state. The desire to prevail, the aggressive charge which made survival of the human species possible during the longest period of its history of millions of years, all these are still in man even though hidden in the subconscious or sublimated in noble and selfless undertakings.

There is no doubt, however, that this tendency persists, and may even be at the basis of stereotypes of neurotic behaviour within the logic of a society in which the economic value scale predominates and is propelled by the stimulus of profit or efficiency.

The impossibility of exercising his aggressiveness in primitive, physical ways, leads man to seek riches, power, or, in short, success. But, actually, this kind of quest represents the expression of ancient fears; fear of destitution and defeat, which stem from an even more atavic apprehension with regard to death.

Undoubtedly, society has evolved and transformed itself throughout the ages with Christian morality playing a fundamental role, and the protestant movement being responsible for the upset of pre-existing social ethics that attached greater value to the fruits reaped by labour and economic activity. Later, socialism, attempted to re-establish certain basic goals that had been relegated to another league in a later life. Nevertheless, the success achieved by the capitalist and productivist systems, and widespread affluence, have driven man away from traditional religious ideals, and, in spite of appearances, even social progress is felt more in terms of immediate benefits than as an ideal to be reached.

The old anarchic, socialist and marxist ideals of benefiting from an ever increasing volume of goods with diminishing effort were quite different. It is no longer an authentic ideal of equality which moves the masses, if it inspires them at all, but, rather, the desire to reach purely bourgeois objectives.

We do not wish to discuss whether it is right or wrong for people to want what they do want, but it is wrong to confuse traditional revolutionary ideals, or pretend to be the victims of a system which, in reality, one does not want to change, but rather hope to exploit even more. If the masses do not see this contradiction in terms, it does not concern us; our primary aim is that of pointing out the incoherence of attitudes destined to perpetrate, to the full, that which ought to be overcome. This actually expresses the incapacity of the majority of the people to judge, in the long term, the consequences of an « Everything NOW » approach.

This confusion is the direct consequence of a lifestyle that is developing and, above all, diversifying at an incredibly accelerated rate, which, in turn, makes planning difficult and often obsolete too soon.

We have already referred to planning as a characteristic of recent Soviet history, which, at first, was looked upon with a condescending eye but which is now widely applied in the western world, to such an extent, in fact, that large and medium firms consider it a standard practice. But, unfortunately planning remains technocratic and above all econocentric, which is to say that it is carried out with purely economic aims in mind. This implies that it is a rigidly hierarchical kind of planning.

In a world whose increasing complexity makes it even more difficult for the center to control peripherical areas, it is logical that planning, which is felt mostly as a limit to the individual's freedom, is unwelcome, since the majority of the people cannot participate by expressing opinions and preferences. This is the problem often debated, even if considered utopistic, of direct democracy. But the consequences of the accelerated rate of the change in habits and tendencies are quite another matter. The original enthousiam for scientific progress is presently being substituted by a considerable frustration at having understood the limitations that science is subjected to in relation to more practical applications within a reality that is itself contained inside the realm of concrete environmental extension and man's biological characteristics. This kind of frustration can only lead the soul toward pseudo-religious forms such as yoga, zen, astrology, the hippy phenomenon; when it does not burst into « magic » and the ancestral hold of the very first attitudes that preceded « religion ».

We are therefore faced with a series of intellectually and emotionally regressive phenomena. It proclaims the immaturity of a society that is incapable of exploiting, in a proper measure, our scientific progress, and the abscence of adeguate resolution thought up by the more aware members of the collectivity. From this feeling of mistrust arises the present cult; we have returned to a naturally less refined and tasteful Horatian « carpe diem ».

This philosophy structured around the present revolves itself in a revolt against technocratic planning because it imposes a certain number of choices for the future. Yet mistrust and rebellion are justifiable, inasmuch as the fundamental error does not lie in the planning, but in planning undertaken only on the basis of economic value. It is therefore right to pretend that future planning must be carried out with the evidence provided by cultural and psychological values, otherwise it will never be able to fulfill all the various facets of human necessities.

Planning is needed for any goal that may be desired; from

ecological balance to the control of population growth, to all other most urgent matters. The alternative to a lack of planning is the ultimate destruction of civilization; yet it is necessary to program planning according to man and not to objectives chosen from a detached and faceless high placed council. The formulation of the consequent decisions implies an aware participation on the part of ever wider strata of population, and only through a concerted effort in the key constituency of the problem - education. It would appear that from the above, we have presupposed that the primary function of education be that of making available for each member of the community a form of culture per se. But education must not be conducive to erudition, it must be capable of forming minds which will exploit and be aware of their free will.

Administrations, technocracies and institutions do not require brains that have been processed by education to be unaware of prejudices. The interest of public and private powers lies more in an education system that produces technicians inside an overall conformism dressed with liberal condiments. A single individual, in his original thoughts, is viewed by the Powers as a social enemy.

However, the very fact that these matters are openly debated demonstrates that we are aware of them, even if, undoubtedly, all resolutions will be difficult and will spring from opinion clashes and anthiteses of conflicting standpoints.

Chapter seven

METHODS AND MEANS

Managing an industrial, commercial, or services firm, be it public or private, has come to require, at the present time, a considerable bulk of specific technical knowledge. Furthermore, if we examine in their relationship with the surrounding environment, the sum total of our knowledge of the procedures and the related research is no complex and vast that we may witness, especially in this recent period, the emergence of a new body of disciplines that tend to group together in a truly and specific science of management.

Universities, management schools, special courses, round tables, seminars, conferences have all favoured meetings and discussions; gradually, a language, often of Anglo-Saxon origin, has come into being, and polemics have arisen on the methods and systems that ought to be adopted in modern management.

Recently, scientific research has availed itself of technical means, epitomized, in our field, by the computer: the thinking machine. Naturally, it is history that has made possible the use of these instruments, capable of reducing the time for complicated calculations to the so-called real time. In practice, as soon as the problem is formulated, it is resolved. We are referring to the binary system, to Boole's abstract algebra, a branch of pure mathematics that earns its name from an Englishman, George Boole, who, before anyone else, created a system — which is really the algebra of logic — from which class or proposition algebra is derived.

The most sophisticated computers, may, then, be built; and their workings can be compared, to those of the brain, or, rather, to those which we believe are the working of our cerebral cortex.

From human fingers, the number of which has motivated the employment of a decimal system, man, since the beginning, has made use of various expedients to facilitate calculations; we may find examples in prehistoric cave graffiti, abacuses-still a miraculous machine in Chinese hands. We have referred to fingers to recall the great subdivision between analogical and digital or numerical measures.

As it is widely known, measures can be expressed with a number made up of units; thus we use a number to represent a physical size; the reverse process may also be undertaken. Thus the difference between analogical computers, which represent numbers through physical sizes (e.g. the intensity of an electric current or an engineer's slide rule), and the already mentioned abacus, is the clearest example.

We have, therefore, briefly mentioned the development of a managerial theory to support the practical management of business and the most sophisticated apparatus of which this theory may avail itself. We must, however, emphasize that first of all it uses instruments supplied by other disciplines such as mathematics, physics and electronics. In conclusion, we may affirm that the management science avails itself for both its theoretical formulations and the apparatus it uses, of what has been placed at her disposal by other disciplines. This is not the same as saying that business management cannot be conducted without supportive methods and systems and the employment of sophisticated machinery. Enterprise, in all its aspects is almost as ancient as man himself, and therefore it finds its true foundations in man's nature, or rather it is born out of the first stages of the present kind of culture and civilization.

Hence, enterprise, as an expression of human activity, does not owe that great a debt to other sciences.

On the contrary, it may be said that man, by producing and undertaking commerce, while managing its own enterprises, has from time immemorial always stimulated scientific research; and in this light the debt owed by science to man as the producer and merchant is invaluable. Let us now briefly refer to the theoretical support already mentioned.

The need for indexes is to be found in the great traditions of scientific methods; logically management studies, must also refer to them. Indexes, in fact, present themselves as descriptive elements of the phenomenon we are examining, whether the objective is that of individualizing the clear situations of the phenomenon, or whether we are attempting to discover the hidden aspects that accompany the most evident one. Yet, to arrive at indexes or indications, the phenomenon must be objectivised and expressed in definite terms, and manipulated measurable results are achieved.

We must realize that, in the final analysis, indexes are comparable, or perhaps are signs in their own right, and we must consider them as such in order to use them in accordance with the most modern and advanced techniques as is the common practice in other fields of study.

The signs that come to us directly from within the business concern for instance, the speed of a spinning machine's spindles — may be termed endogenous, while those that come from the market, or, in other words, from the environment in which we operate — for instances, the subscription rate for a specialized magazine may represent a useful index for the management of a specific firm may be called esogenous.

Reaffirming, thus, that collecting facts, if we wish to substitute this term for the word index, tend to endow the object of our attention with an appropriate form for subsequent analyses adn elaborations, and this allows us to talk of mathematical patterns to be used for a verification of the variables we have assumed by inserting them in an adequate pattern. Naturally, this mathematical pattern, or model, must be as faithful as possible to reality; unfortunately, reality is so complex and the interplaying factors are so numerous that even the most complicated pattern cannot be other than a glaring simplification of that reality.

One of the most important characteristics of a mathematical pattern is that it is useful in a multiplicity of practical instances. This justifies our method, which we may, perhaps, term a research expedient, without taking anything away from the specific values it possesses.

The relationship between all the different factors that determine an economic or socio-economic situation and the possibility of representing the situation in a synthesized form, giving way to a logical or mathematical process that attempts to place in evidence the essential relevant elements of the matters examined. Clearly, it is a process of abstracting and organizing these fundamental facts. Consequently, the formulation of mental or mathematical models finds specific relevance when it is applicable to more concrete instances, when they acquire a degree of proportional convenience in relation to their extensibility to phenomena that belong to different situations or, perhaps, even to different fields of study.

Let us now consider that many factors belonging to an economic scheme cannot be translated in numerical terms and cannot, therefore, be included in mathematical elaborations. We are referring, for instance, to psychological evaluations, behavioural conduct, to the so called « social indications ». We assume that non parametral methods and non-numerical calculating procedures are being studied and that research has been undertaken for new, more appropriate algorithms capable of responding more directly to fields of enquiry such as the ones we have already mentioned, in relation to psychological and behavioural sciences, that will gradually acquire greater importance in the business world.

In everyday life, we constantly experience banal and hurried evaluations that are determined by what we call our « impression ». We may, for example, judge an individual interested or involved in varying degrees according to varied fields; we can feel a given temperature more or less comfortable or pleasant. These evaluations, especially if broadened to wider confines, will provide a useful and important source for comparison. The deepening of this kind of problem must be regarded with the utmost consideration, irrespective of the solutions one may reach. Clearly, the difficulties will increase as the respective group or context is extended.

The vital place attained by psychology in relation to motivations and learning translates the interest in studies onto a plane where man is the leading character. But the effort expended in resolving industrial problems, or in finding new instruments capable of facilitating the decisive choices, leads to an acceptance of solutions grounded on statistical facts.

In this instance, we are not referring to statistics as a instrument for pure research; as a matter of fact, when we refer to certain data and informations, we have obviously already availed ourselves of statistical methods and techniques. Our intention, on the other hand, is to set it against other working formulae, like those based on pattern, or models, the practical observation of events as they happen, so that we may arrive at statistically valid conclusions, which — as such — may be taken without further elaboration as fundamental bases for decisions.

When we are concerned with data that is external to the firm, the problem of the vast quantity of information on which we may rely — or rather that loom over us as we attempt to extract what is relevant to us — arises. Their accumulation is such that it cramps the sphere of an action in sifting it all out.

Thus the need for the development of automation in the research of informations, in order to provide scholars and businessmen with data that may be of some use to them. This technique is a prerequisite to the progress of scientific disciplines which can find in it its own methods and systems.

Through the ages, these methods have been organized inside systems which, by applying more up-to-date equipment, reap quite considerable results.

From the most rudimentary techniques (we are tempted to call them almost « instinctive ») adopted in classifying files of a small library, we can arrive at the proper management of an enormous quantity of data. Through the mathematical instrument one may translate languages ad even create artificial languages as the ALGOL has done for programming.

We would like to point out the fact that even small and medium size factories feel the consequences of the aforementioned problem directly.

As a matter of fact, in many instances, a possible lack of data regarding the factory itself, or the difficulties encountered in digging out information on external factors are only marginal when compared with the organic conservation of said data, according to a system that need to be complex but may be only relatively efficient, in relation to the dimensions of the concern taken into consideration.

Small and medium size concerns possess archives that, more often than not, are kept in worse tended storerooms than those used for raw materials, reserves and deposits. These are often emergency sites, in which tons upon tons of paper lie dormant and unused: they are difficult to reach and are classified according to an original, elementary system the logic of which has been destroyed as a result of having been retouched through unthorough and half-hearted modifications. Instead of sorting out only the important documents, which possess a statistical or historical value and may be useful references giving perspective on preventive calculations, all is kept irrespectively of its value in a wholly chaotic environment. After a few years the acquisition of data relative to the factory becomes a truly impossible task.

Nobody wants to throw away written papers, yet is necessary to preserve only a relatively small amount of relevant documents. To keep and care for everything may lead to total disaster.

Often, the courage to send material to the retting-pit fails, because of the fear that in the pile there may still be some useful isolated paper or document, and thus the situations is willingly left to worsen beyond repair. On the other hand, important documents are often destroyed immediately; for instance the ledgers, which would represent a useful source of reference for an over-all picture of the undertaking's management record, are destroyed in the fear that they may ultimately contradict the data contained in the official book keeping.

But let us return to our fragmentary and all-too-swift list of comforting aids for industrial management.

The individual or organized team of experts charged with analysing the different systems fulfill an invaluable function: that of actively managing the informational flux. In order to do so the employee requires a precise knowledge of the aims proposed and the necessary ability and competence to discern in time the specific problem in a global overview of the firm in the precise moment in time.

The firm's organism to which these informations ought to be channelled must be located properly and this should reflect the interests of the receiver in relation to his own sphere of action and to the kind of business process over which it presides. A deep awareness of the various jurisdictions and responsibilities, the degree of discretion of individuals to whom these information are sent, are taken for granted.

This, in turn, leads to an overview of the firm's conformity and its procedures, in order to smooth out the functioning information-system. If it is properly carried out, the firm's management will be relieved of a great deal of work.

Clearly, the risk of disorder caused by evaluational mistakes must not be overlooked and we shall return to this question. However, the risk is inevitable; it is inextricable from any situation, an intrinsic part of life in general and of enterprise in particular. Nevertheless, we do not believe that the efforts, studies and research conducted to objectify judgements and evaluations may ever extinguish or reduce the influence of the ineluctably subjective evaluations on a great deal of factual data.

While our concern is with information, it appears relevant

to stress a certain category of facts. Today, almost all industries possess automatic devices connected to machinery which determine their self-regulation.

The passage from manual equipment to that which employs energy foreign to man's physical capabilities, represents the dawning threshold of the machine era. For a long time, machines have been operated by men, even if gradually they have acquired a certain autonomy in their being able to undertake a diversity of operations and adapt to a multiplicity of requirements. Now industry can rely on installations that are practically able to make decisions, to work, stop, change, effect operational variations, according a preordained plan.

It has become clear that for the regulating instruments of most disparate kinds, according to the type of industry or technology by which they are employed, there are often corresponding graphic control schemes that fulfill the duty of these regulating instruments, but that — obviously — provide very important data for judgements as to the work undertaken and the degree of efficiency achieved through it.

Yet, the automatically-achieved essential data do not accrue only within industrial precincts. In the field of transport and distribution I have personally witnessed the usefulness of tachigraphers linked to motor vehicles. A paper disc inserted daily in the tachigrapher provides data on the vehicle's and the driver's pauses, since the graph drawn on the disc indicates minute by minute the behaviour of the machine, hence of the driver. In relation to distribution with the door to door selling system, this instrument will inform us of the time taken by the transfer and the length of time expended during the visit to the client. Significant information may, after a little practice, be obtained in a flash.

Furthermore, together with controlling functions, this method is useful to predetermine distribution itineraries. By recording the real and actual conditions in which distribution takes place, judgemental errors, which the most accurate accounting of kilometric distances by other theoretical informations cannot alleviate — may be avoided. It is evident that the tachigraph disc informs us of the conditions of the itinerary and of the complications derived by traffic contingencies, and the slowing down and breaking periods are recorded with the utmost accuracy.

Furthermore, as we have already mentioned, the graph's course gives us a more precise indication of the driving ability of the employee. We therefore wish to emphasize the validity and importance of automatically transmitted data; even if of an analogical kind, they often provide the quickest and sharpest interpretations of facts and circumstances.

Perhaps business techniques may also lend themselves to methods of a medical nature. The bioelectric phenomena of man are studied with the graphic system: electrocardiograms, electroencephalograms, pneumograms, dynamograms, etc. Undoubtelly, the business' activity may be followed by increasing the reliance on analogical methods; contrary to what may appear at first sight, immediate simplification would came into effect. The technical means, as we have noted earlier, exist already: perhaps in the business field, or rather in the field of business management, its full advantages are not employed.

We hope to have recalled the methods and instruments inherent in the systematics of business research, as it has developed through time and particularly in the Anglo-Saxon world during the past few decades, and more recently in our country.

The special contribution, different in many qualitative aspects, provided by Italian researchers, must be noted; we propose to return to them for an analysis of what we deem to be their particular characteristics.

We have talked in general terms of a decision-orientated research; in closing this chapter, we wish to underline what is known as O.R. (Operative Research) inasmuch as this system is characterized by the effort to adjust the model with reality and formulate models that may be useful in more than one instance.

Chapter eight

SYSTEMS, MAN AND ECONOMIC DYNAMISM

MAN AS THE DETERMINING ELEMENT IN BUSINESS MANAGEMENT

So far we have discussed matters concerned with informations, mental models, decisions; we must now pause on the most prominent feature in our survey: man. We must take into account the individual's behaviour within society, in order to understand the latter's influence on the former. Without exceeding the boundaries of this work, we would like to examine the individual by starting with elements more directly related to the thread we have so far followed.

Man's brain is a reception centre of external stimuli that correspond to orders for actions in reply. Having introduced the concept of command, we enter the world of cybernetics, a science that essentially studies regulation methods of living being's organs and instruments or machines constructed by man. In the first instance it will revolve around an analysis of biological process, in the second of physical, mathematical and electrical processes.

Evidently, while discussing regulating methods, we are also referring to transmission of orders, stimuli and informations.

From stimuli we may pass onto perception as the fruition of an internal elaboration of the stimuli through the intervention of individual experience. The dimensions in which the individual places himself — space and time — are the outgrowth of experience. But experience rests on the faculty of memory for self-preservation. It is necessary to introduce here the concept of intelligence in order to arrive at the concept of human learning. Intelligence has been defined as the capacity to resolve problems, or as the necessity to make good use of the past experience. There are some who consider intelligence as a global faculty of the individual, and some who refer to it as a whole comprised of particular faculties.

Memory and intelligence lead us to learning, which nearly always proceeds through several attempts and failures: this is the adaptation process, a parallel of which — by antithesis — is learning through conditioned reflexes. Man's behaviour, then, is determined not only by intellectual learning but also by emotional life.

Freud in his « repeated coercion » theory expresses the belief that childhood emotions and experiences are sovereign and ever recurring in one's life.

Others rate those emotions that manifest themselves in certain, given circumstances as being more important. Nevertheless, there is no disagreement on the importance of the subconscious, even if procedural methods arise out of different conceptions and hence follow varied paths.

At this stage, let us examine what is determinant in the various ages and circumstances. We may make a distinction between the gratification of primary needs; the action that reaches beyond an immediate aim; the necessity to mantain a sufficiently sharp self-image. These three categories of motivations relate to the three ages of man or humanity: childhood, maturity, old age. Yet they also refer to three degrees in the human condition, the first of which is determined by the quenching of fundamental needs and the others arrive at more interior, spiritual requirements.

We have listed a number of concepts that cannot but reinforce our statement that man as an individual differentiates himself from all other human beings. On the strictly genetic and organic planes, this finds more, unquestionable support.

Let us add that each person's actions are determined by more general factors, such as may be culture and the characteristics pertaining to the group in which one lives; which is true in all form of communities. Absolute values, then, are false and misleading, and cultural, moral or other such values must be considered only in relation to a given context.

Each man is the result of a complexity of social and individual factors that each preference and thus each choice he makes is a wholly personal matter.

If we were called upon to determine the number of individual characteristics, we would be forced to list all the determining factors that contribute to the formation of an individual: the possible permutations contained in said number would be represented by a factor multiplied by N where N tends towards infinity.

Picking up our main thread, once again, we must conclude from what we have examined, that the various methods, instruments and systems in a global management will never be able to determine choices that are valuable in absolute terms for all concrete examples.

The bearing provided by the individual or group determines ineluctably and characterizes the formulation of the decision.

To clarify what we have felt necessary to express in relation to the working processes of computers and the human mind, and in order to point to their correlatives and differences, we will limit ourselves to a brief consideration. The so-called thinking machines, are little but extensions of the brain, in the same way that cars or telephones are extensions of physical human activities; they, in other words, stretch the capacity of man for covering physical distances at a certain speed and the faculty of communicating and transmitting his voice over wide expanses of land. But computers are constructed by man and as such, like all other machines, are tied to the programmes fed to them by man. Each choice is taken in view of the original facts that have been programmed; this is particularly obvious when the computer is fed with estimated data rather than numerical measures, in a physical sense.

The innumerable calculations that the computer provides us with, are originated by a long series of estimations conducted by man. It has been rightly postulated that the computer is not superior to its programme, which signs in sum signifies that it is much more difficult to obtain the original information than its elaboration. The question asked, if incorrect, cannot be provided with a useful answer.

Clearly, the correct question can only be fed by man, while the logical process that follows may be carried out by machine.

When we refer to man, it is evident that we are, in a concrete case, referring to the individual or the group concerned.

But the single man or men cannot express a valid judgement or evaluation for a wide crosse section of individuals. Each man cannot speak for the community; his judgement, even if it is determined by his adherence to a particular group, will always remain subjective and as such it will never be able to reflect a collective evaluation. An optimal decision is unobtainable, only an individual decision, or one arising out of an intellectual compromise among several persons on a certain, specific situation in time, and according to a host of subjective starting points, may be reached. Yet, we must not forget that no fact or evaluation may exist outside an external cultural context, since it refers itself to a freely or forcedly accepted scheme of social behaviour, which cannot be ignored by the individual who is called upon to make the ultimate decision.

This supports our theory on the determining influence cultural mutations have over all other decision and consequentely on the enterprises' life and in the final analysis on economic dynamism. Whether the subject revolves around micro or macro economics, man remains the unquestionable protagonist of each decision and he cannot delegate this responsability to formulae, models and computers, because he is both the active element and the object of industrial and social choices within the precincts of time and space in which he operates.

This is how we perceive the factor — man — in a business and therefore in an economic context.

We have already raised doubts on the values inherent in the adoption of given systems in management; we shall clarify these apprehensions later, but for the time being, we wish to affirm that methods, instruments and systems have a wider role to play in the formulation of plans directed towards the realization of socio-economic development in a community.

This has been placed in evidence in the opening part of this work. Here we wish clarify the fundamental motive for the differences in approach.

Choosing a plan or programme, destined to influence the social development of a large community, finds its presupposition in the political choices arrived at by the community or in the kind of governing body it is existing under. The directive lines of the plan are therefore constructed according to the sum of values held within a certain political programme. The adherence to a concrete political configuration arrived at through a natural ideological evolution conducted by a handful of forerunners of the prevaricative imposition of a single man or group is already present. The democratic development, the ideals championed by an élite, the will of a group who has seized power, lead to an organization of society according to the finalities that constitute the objective of a people, an élite, or in any case, a dictatorship.

Clearly, the plan will be none other than a mean, irrespectively of its complexity or elaboration, to actualize choices that have long since been made.

A business concern is a very small organism and, since it represents a single entity in a extremely wide context, it cannot be motivated by such explicit reasons in its choices. A condition for survival in business is to achieve positive results through efficiency, and under any political regime. Evidently, such a goal does not constitute a valid indication for choices and decisions. A business concern operates in correlation with other businesses and each must aim at acceptable results.

In contrast with the clauses applicable to businesses the programmes destined for the community follow quantifyiable goals evaluable only according to ideological forewords. Enterprises' dynamicism is conditioned by political factors and its action may be tied to predetermined programmes and plans, but within enterprise such factors are added to other evaluative criteria and do not constitute in themselves neither a simplification nor a sufficient indication.

A comparison between programmes and business planning obliges one to move along a badly defined ground because along with small concerns we are confronted with huge complexes like the multinational companies, that may avail themselves of more powerful means than those available to States. For them, large plans are indispensable, since they tend to undertake situations and provoke repercussions that require political intervention. The immediate aim of their operations is not limited to profits; it is often geared to seize more power; from these vantage points they are able to reach enormous profits that have little to do with actual management.

We can therefore state that the great enterprises are foreign to the context we have so far explored, because we have been concerned with the rule rather than the exception.

Huge enterprises, after having reached a certain stage, live and prosper according to perspectives that are more akin to those conducted by the colonial nations, than to an economic logic.

Socio-economic plans and programmes and business plans and

programmes may both be subdivided according to short or long term considerations. Short term considerations may be represented by those in which an organization must vary its methods of employing certain structures; long term, those in which the organization must change its structures.

If mutations in the environment, in resources, and in technology gather greater momentum, the present organizational structures, of a State or a business, will soon become obsolete. In such a situation, what is defined as being long term, will become increasingly more immediate because of the necessity to change the already inadequate structures to meet each new situation. If it does not happen, or the transformation cannot take place, the organization will loose in efficiency. If the change is much faster, it will bring about an emergency situation: the organization will enter a crisis and consequently it will not keep pace with the economic dynamism. The alternation between intervention on the uses made of the structures and the transformation of said structures, follows the changes in a cycle that continually gathers momentum.

In this realization we find a premonitory sign for the necessity of a programming at a high level, aimed at establishing, on the basis of fairly loose predicaments, the great social objectives to be reached. Only in this way, the disfunctions and the diseconomies that are produced by an uncontrolled development and determined by the fractioning of programmed goals on the level of single factories or as consequence of the actions undertaken by large conglomerates of the gigantic multinational groups, may be avoided.

We believe that the market's function has been surpassed, but, as we have sustained elsewhere, the important choices relative to the commmunity must be taken by the government, in a system in which a priority scale redimensions the economic components before they affect the market.

Often, when talking about the sort of theme we have encountered, one meets terms like strategy and tactics.

Evidently, these terms are employed in a figurative sense; however, on the basis of the premises outlined at the onset, it should be clear that we regard strategy, in the great social organizations, as revealing itself in the political formulation of goals, and tactics as inherent in the preparation and realization of the consequent plans and programmes. In a business concern programming is strategy and the operational phase is tactics containing (at the present time) the correction of the course followed when changing situations emerge.

This should occur without altering the objectives: to modify them, at least the most important ones, comes under the heading, as we shall see, of business strategy.

DIMENSIONS AND DEVELOPMENT OF THE ENTERPRISE

The modern tendency towards an ever increasing size brings us to consider whether the dimension of an enterprise has an intrinsic bearing on its efficiency or not.

We think that increasing the size does not necessarily imply an improvement conducive to optimal conditions. Many variables determine in a given context the most appropriate economic dimensions of an enterprise. This is determined by the best use of resources and production factors for the ends of the enterprise, and for the society within which it operates.

There is therefore a definite distinction between development and dimension, the latter being part of the dynamics of the enterprise and the former a mere fact. There is furthermore a distinction between stressing technology (giant industrial complexes) or management (organization of sales, purchases, and financial activities). These two kinds of business economics do not necessarily go together but they are almost invariably found so. Technological gigantism has been studied especially through the theory of costs. Less research has gone into the management economics of the huge business concerns which rely on organization of research, vast commercial networks and continual diversification.

We would like to contrast the above dimension economics with development economics, in which regardless of size all latent possibilities already in the enterprise are searched and exploited.

Mere size cannot guarantee an advantage in terms of efficiency or economic viability. Disadvantages are inherent in large size which reduce the economic benefit derived therefrom.

When giant size is the means to other ends the problem must be tackled on another level, since evidently profits are not measured in purely economic terms. We must therefore briefly refer to mamouth multinational companies.

An audacious parallel between the joint-stock company at its birth (attributed traditionally to 1602, the year in which the Dutch Company of Oriental Indies came into existence) and that which we find today: the multionational — as an instrument of a mature capitalism — has already been drawn.

In the former we find the typical characteristics of the modern joint-stock company: the pro-quota participation, the free currency of the shares, limited liability, and above all the vocational practice of macromanagerial activities.

In a historical moment in which the newly born capitalism nourished by protestant ethics, found the most fertile ground, in explorations and discoveries, the Dutch Co. of Oriental Indies began an immense project of penetration and conquest embracing wide territories, from the Persian Gulf to Indonesia, wich involved the collection of finances from the savings market, the employment of fleets, armies, civil and judiciary administrations. The Company had invested upon itself sovereign powers, directly derived from the Prince, that in so doing created a veritable State within the State; she was the instrument apt to mobilize private resources at the Crown's service, but she also was a considerable centre of economic and political power.

The great joint-stock company of our time does not employ fire and steel in its operative conquests; she avails herself of more cunning and perfect penetration instruments to arrive at substancially analogous results. The Company had divided in the same number of monocultures the Regions of Indonesia so that she was not forced to diminish the prices of spices on the Amsterdam market. If the local farmers began other cultivations, these were inflexibly destroyed, since the requirements of the autochthonous populations were completely irrelevant for the Company.

Today, even without sovereign prerogatives, the multinational companies influence economics and the interest of whole communities and of free and independent sovereign political systems: for an all embracing example, we may think of the events that developed around certain North American companies, in the Central or Southern Americas, and those connected with the renowned *Seven Sisters* oil company.

The historical parallel between the newly born joint-stock company, and that of the present day, demonstrates in both the permanent intermingling of public and private interest and interferences, the dissociation between ownership and control, between share holders and administrator. The history of the joint-stock company is the history of capitalism, from the industrial revolution to the spectacular progress achieved in the last few decades, but it is also the history of the exploitation of man by man, of speculations, crises, and the most cynical and thoughtless manoeuvres. The collapse of John Law's Bank and his Indian Company, threw France in a crisis that lasted over a decade. The colossal fiddles operated by numerous joint-stock companies in England, that culminated with the 1720 Bubble Act, were the undetected forerunners of the great 1929 slump, to which the irresponsible policies adopted by Banks and Investment Trusts contributed a great deal.

Today, declaring wars, forming alliances and minting money are prerogatives of the States: yet, how many times have these powers been exerted under the incentive and according to the interests of these financial giants.

Even today, in a modern joint-stock company, ownership and control, power and risk are not necessarily joined; their dissociation had already been denounced by Karl Marx.

The extreme diversification of share holders ownership reduces the members to mere sources of capital; assemblies — in theory sovereign — are deprived of authority by the Administrative Council which is dominated at first by the majority, then by insidious and powerful minority groups; profits unknown to ordinary members are ploughed back lock stock and barrel in the company without notice; always growing and often international groups come into existence and multinational companies become the protagonists on the European and world scene. The relationship between business and society is characterized by different shades, according to social, political, economic peculiarities of the various nations.

In the U.S.A. the 500 major enterprises of the country embrace some to thirds of the extra agricultural activity. As Chajes says « The great enterprise is the dominating private institution of modern everyday life in the United States ».

The progress of progressive institutionalization of large enterprises, as Cottino points out, is an irrefutable aspect of our social reality; what cannot be accepted is his conceptual and philosophical justification.

Through motivational research and hidden persuaders, one tries to identify — in the consumer's mind — every producing enterprise with the values most honoured by society: uprightness, reliability, security, fatherly love, childlike innocence, neighbourlness, sociableness, etc.

The times of the great Companies and the great capitalists of the nineteenth century — Rockefeller, Vanderbilt — seem so far removed from today. The truth is that if society has to limit herself it means that she does not, in practice, have any limitations; her politics influences public interests, prices, dividends, incomes, currency, exchangers stability.

The multinational follows industrial objectives in a context that has nothing more in common with that of national enterprise. The huge capital available to them may then be placed in the most different nations; the choice rests on political considerations rather than strictly financial and economic ones. The political stability of a country, the certainty that production will not be halted by strikes, will decide the location for the installations and the entity of investments.

Customs and excise duty, as well as fiscal policies entertained by given nation are also important for the flux of capital that the multinational transfer with an ease that astonishes even the most alert central Banks.

The characteristic aspect of management in a large multinational undertaking is constituted by the availability of numerous instruments for the realization of transfers in financial flux.

The finance obtained in a given country will be exploited through a branch in a different country, or deposited in the central house, thus escaping all legal restrictions.

This may take place through interest, services or commission payments between the various branches or associates, or through the subscription of shares placed onto the market by a company member of the group. It is particularly important to state whether it is possible to alter retail prices of goods or services between the various companies in the group, so that the profits achieved are distributed in the most sensible manner.

Enterprise aims to minimize customs and excise costs by transferring the profits from the countries enforcing the highest rates to those that apply lower ones.

Thus, a management that is able to exploit the possibility of working in all four corners of the world reaps the benefits of the advantages we have just pointed out. To this extremely favourable condition we must add the possibility of availing oneself of a thorough net of information, experiences and alternatives, also on a plane of continuous exchange between the men and methods of the different countries. We could therefore say that never as much as in these cases the opportunity to manoeuvre large sums of capital creates power, and the power obtained, in turn creates the opportunity for gigantic profits that have nothing to do with the purely economic validity of said enterprise.

Omnipotence has innumerable and unpredictable ways of operating and influencing external operations; the support obtained by expansionist policies, the reassurances of sovereign powers, a delegation to a select number of managers chosen to take decisions that will affect the whole community.

Before we conclude this brief examination it is worthwhile pointing ou that, in the United States, the existence of an internal market of gigantic dimensions represents a situation that has consolidated itself through time, which has had the effect of justifying, the creation of always bigger industrial and commercial complexes.

At the other end of the scale we have the European situation, which for reasons that are too obvious to be dwelled upon, and even after the constitutions of the Common Market is behind the times and is conditioned by enormous obstacles of a nationalistic nature. It is therefore possible to hold the most diverse and totally conflicting attitudes with regard to a share company; agnosticism, however, is neither justified nor justifiable.

By accepting the large companies as they are today, we accept the inevitable consequence of the radicalization of the extreme aspects of the present situation which will lead to an unpredictable and uncontrollable situation that may then branch into any one of the innumerable outlets available.

CHANGES AT THE ORIGIN OF ECONOMIC DYNAMISM

No change may insert itself in any branch of human activity if it does not find the necessary conditions that will allow it to weigh efficaciously on the preexisting situation. No new idea or scientific discovery or technical innovation may spark off an evolutionary process if it develops accidentally. A flash of lightning in the dead of night cannot represent a guiding beam in the dark. Many intuitions and inventions, throughout the history of man, have not been able to reap fruits because of the absence of unreplaceable complements. We only need to remind ourselves of the enormous quantity of genial projects documented by Leonardo in his writings; they were not used because the condition of other techniques had not reached such degree of perfection.

Through antiquity, among certain tribes, some methods employed as substitutes for human activity, in the form of artificial energy to replace man's physical endeavours, have been found; yet they were never exploited fully since they lacked the knowledge pertaining to the science of mechanics. New methods may, on the other hand, be introduced only when a situation that requires a collective solution comes into being, and in this eventuality, it will arise out of the necessity to discover alternatives to the old and obviously, in some way, ineffective systems. But when alternative proves necessary, when the obstacles in its discovery are not too great, or there are no conflicting interests stubbornly defended by whoever holds the reins of power, in its path, then the emergence of new methods and techniques, of those expedients vital for the development of the human society — in short, inventions — begin to come to the surface.

The changes occur along the road followed by society, at some stage slowly, across eras that appear crystallized in their traditions and institutions, and at some other stage quickly. In the past few years, we have witnessed profound transformations in all fields, taking place with an unprecedented rhythm.

Change is a constant and in our field it determines the dynamics of economics, irrespectively of whether we pause on the economic aspect of history as a whole, or whether we look at the great organizations and businesses. And enterprise is the object where changes take place in all their practical reflexes, touching men and things in everyday existence, in the ultimate and most authentic expression of the tendencies that regulate society and economics in general.

Today, more than ever, enterprise suffers continual changes that may be verified in all fields, so much so that we may speak of a real emergency with regard to enterprise, because the difficulties that constantly bar its path are truly great. Man's nature, his mind, are not made to take part in such a stormy evolution. From here, a crisis related to the factor — man — within the bowels of enterprise on any level it operates. But above all, at the decisional level, in the practical management of businesses, the swift transformations that upset each aspect of life and each human relationship, produce problems, for those involved to find an orientation and direct the concern and the employees, according to an internal order to be used as a yardstick. Even at a decisional level, man is tied to behavioural stereotypes and cannot modify them with excessive speed without paying the price of dangerous internal upsets.

From here the anxious search for systems, formulae, elaborators who may provide intellectually justifyiable solutions, which — in a certain sense — relieve management of some responsabilities that do not find gratification in precise fundamental directions.

Almost always, those are problems with too complex resolutions, because the comfort of traditional calculations lose their impact and it proves increasingly difficult to arrive at them through numerical elaborations and models constructed on previous experiences.

As we have already mentioned, methods, systems and models for management will have decreasing value in the direct and practical applications to specific instances, in the measure in which certain factors, that are not strictly of an economic nature, come into play.

Let us now examine another determining force for economic dynamism: reticence toward change. It is an element of the utmost importance and it must also be considered as a constant of development.

This reticence manifests itself in the most diverse forms; yet it always derives from a self preservation instinct, from a desire to protect a certain existence, a traditional way of life which has matured into an image of oneself and of the surrounding world, in which onc can feel at ease and identify oneself with one's environment.

The skilled worker, the foreman, are almost always unfavourable to changes in machinery, even if this leads to greater efficiency and requires little added knowledge to operate. They want to preserve present machinery in an understandable self-preservation instinct. Often, the manager of an administrative office does not wish to adopt new procedures, even if they cut down on work and responsibility. This kind of reticence is present in almost all sectors, it does not depend on matters related to efficiency; its prerogative to disrupt what have come to be everyday habits is sufficient to arouse mistrust and apprehension. Naturally, this is not a general attitude, but especially on the threshold of a transitional period from one system to the next, the number of those who, in a given environment, accept novelties with a truly favourable spirit are certainly a minority.

The majority still feels tied to instruments and systems from which they have already received some gratifications or a sense of security. The identification with a kind of life, the moulding into a gerarchical system, the sense of belonging to a determined category, may be threatened by a modification, and this is what really determines resilience or aversion toward it. But an individual's problems are often similar to those of his firm that is likewise slow adapt to changes and their consequences.

The knowledge of the sacrifices that the community has always suffered during transition periods in its structures, having to undergo deep crises, revolutions, wars, explains why all major, significant changes take place with difficulty, uncertainties and violent contrasts. Changes may originate from variations in one or more of the classic factors singled out by traditional economics: resources, work and capital.

Resources produces repercussions, during their retrieval when they are used in large quantities and when they become scarce. They determine the business' dynamics in the phase in which substitutes are pursued, as it is now happening in the effort undertaken by public and private industries with regard to alternative sources of energy.

Work, understood as a production factor, feels the consequences of new acquisitions, not only those inherent in the social progress, but also under the profile of a full knowledge of man in the biological and physiological fields; from here the importance that ethology is assuming in the sphere of research. The concept of human work and its place in a scale of values suffers from innovations in both the technological and ideological fields. A different consideration of its importance in economic factors is indebted to the organization of the working classes under the drive of which we have arrived at laws that reflect social demands.

Capital, or rather movable goods consisting, to begin with, of suitable objects or materials and later of money, continues to develop more and more and achieves always greater importance until it forms a power that substitutes other kind of power; in its concrete expression as money, it determines prices, interest rates and financial phenomena of all kinds and it constitutes the passage from a primitive economy to another which is more closely related to the true dawn of capitalism. It determines the advent of urban civilization, the creation of class distinctions and it controls the centralized structures of the State.

An incalculable number of subsidiary factors come into play with the ones mentioned, in a constant process of interaction, so that each factor is, in a certain way, cause, and in another, effect, in the variation experienced by the other factors.

Saint Thomas of Aquinas, Malthus, David Ricardo, Proudhon, Adam Smith, Friedrich List, Warras, Keynes have proposed theories on savings that have had a deep effect on the economic world and hence on the policies followed by enterprises.

Ideologies and ideals, always appear critical of the situations they develop out of; they criticize morals, cultures, social relationships, power. Often these ideologies, if they insert themselves in a ripe enough situation, produce revolutions of all kinds; but, a general characteristic is that in the creators of ideologies at the basis of a revolution, belong to a social class, group or profession that has little to gain from a turnabout of situations. We could say that the great social uphevals have been fathered and supported by elements who were often part of a social strata most favourable to a preservation of the status quo.

Often scientific knowledge, culture, a sensibility towards general problems produce, in the economic field — on the part of managing directors — initiatives that lead enterprises or great organisms to create new structures which without being pressurized by contemporary events, attempt to predict the future. In such a manner, and especially in the industrial sphere, one exacises considerable pressure on the environment.

What it boils down to is not a form of adjustment to the circumstances, but, rather, an active attitude as the fruit of competition in the realm on which enterprise and to a lesser degree the great organisms and only in certain aspects the States must operate.

The adjustement of the structures is, then, a strategic expedient in the conflictual situation in which the organization finds itself in respect of the others that operate in the same area. From here the importance of the role played by predictions and in all other implications that it contains in terms of research, its economic cost, the ability and intuition with which it is managed.

The public or private entrepreneur will have to verify the possibility of adjustment of the internal structures to the external changes. He will have to take into account the time most suitable for these adjustments; from here the importance that today more than ever the flexibility of management and the concern's structures, has acquired.

The speed with which readjustment takes place thus gives an indication of efficiency.

Admitting that in long term considerations the continuity of situations is more theoretic then actual, we consider the mutation process as a normal occurence, and this « normality » may be counterbalanced by a swifter rate of change. When the rhythm becomes extremely fast or other complications and arrhythmies, that transform the economic dynamism in a completely atypic movement, come into play, we define what arises as an emergency situation.

Thus, emergency in this context, signifies anomalous even if we cannot term them exceptional situations; the same applies for microeconomic matters as well as enterprise economics.

Let us now look at some practical circumstances that may be defined as emergencies. When a concern finds itself with equipment that has all of a sudden gone out of date due to the unexpected and swift development of new production techniques; when the distribution system on which rests the commercial activity of a concern cannot keep pace with the market's exigencies; when an industry witnesses the daily worsening of the economic conditions surrounding the availability of its most needed raw materials; if, when faced by one of these situations one lacks the means and capacity to implement a radical transformation of the structures, then, one is confronted with a true emergency.

To conceive of an emergency as an exceptional circumstance is wrong, inasmuch as what has begun to happen will continue to happen with greater intensity and impact in the future. Tomorrow's economy is, predictably, an emergency economics, for which we must be prepared. In order to do so, we must think in terms of temporality and risk. These words are destined to cross the threshold that separates private from public enterprise. Soon these distinctions will cease to have meaning and the public sector will have to come out in the open, leaving aside its priviledges and enter into competition in the struggle to reach optimal solutions, by adapting to constant transformation as a way of life and to flexibility as the only option open for survival.

Everyone must be convinced of this necessity; today too many people believe in the durability of two distinct spheres: on the one hand that which fosters public enterprise and services, on the other that in which private undertakings operate. The fact that we rely on the latter to cover the losses incurred by the former, must not lead into the false assumption that it can subsist forever. In spite of everything, the awareness of the limitations within such system is growing. Only a truly productive work, to be understood as general praxis, may allow public and private enterprises to live together. Similarly, only a great efficiency in the public sector may widen its sphere of action. To increase the number of public employees does not go hand in hand with an improvement of the organization of public services or social welfare. The public sector is already in an emergency situation; we only need to tink of local corporations, their deficits, their disfunctions. This in *the emergency* that requires immediate, highly qualified intervention.

But, returning to a more detached position, we would like to point out that the abnormalities of objective situations in which all businesses find themselves working in, constitutes very particular forms of management.

During a war, enterprise undertakes all measures except for a slackening of production; in fact, it stretches production to its utmost and special conditions, greatly encourage research for new techniques, new devices and more efficient systems of work. Wars, or the preparation of weapons, are sources for discoveries that will later be applicable to the civilian world. But wartime management is such that it may, under all respects, be regarded as emergency management: due to exceptional finance, working conditions and to exploitment of alternative raw materials. The market provided is wholly unique, and this is also valid for strictly non warfare articles (in short, for all items produced in such a particular time).

Emergency sources for enterprise may be those natural calamities that, especially in the past, have influenced the history of humanity and any other analogous event which has disrupted what — at that particolar moment in time — was the habitual « way of life ».

As a final example of emergency economy we would like to mention the situation that has developed in the last few decades, following the creation and development of the State of Israel. The problem has been so successfully analyzed by Facchini e Pancera in their work « Dipendenza economica e sviluppo capitalistico in Israele » with the aid of such wealth of facts to make their examination both fascinating and very actual. Their treatment develops in such a tight manner, following facts and events, that it appears impossible to quote any one passage from it, also due to the fact that the work must be apprehended in its entirety.

We must realize that the facts inherent in the situation are too contemporary to allow an objective interpretation, and the approach followed by those authors, as they themselves admit, feels the influence of a critical attitude.

As far as we are concerned, all too often the attitude and economic policies of the Israeli government have been the source for qualifying attempts, while it is clear than a certain line of action — even an economic one — is determined by the particular situations in which this nation has found itself, even before its actual constitution as a State.

The book is rich in objective and fascinating references related to the nature and diversity of its territory. Water resources are found far away from the fertile areas and hence the immense economic expenditure incurred for irrigation. Their examination of the renowned kibbutzim, their reason for being, the description of their functions, the economic results also achieved is also remarkable: here we have a supreme example of emergency undertakings.

The industrial sector is no exception in its genesis, the type of finance available, the political intervention in production carried out by the government, the Jewish Agency, the Histadruth in the overall atypical conditions for existence. A noticeable characteristic is the coexistence of private operators — for the main part small entrepreneurs — at the head of personal or handiwork businesses. The great private enterprises are, on the other hand, in the main, branches of international companies.

The internal market for production has obstructed the creation of large industries and impeded the advantages obtainable through large scale economy. In more recent times, and in conjuction with strictly wartime conditions, wider range industries have developed to support the increasing demand for military equipment. The description of the distribution policy of the population across the whole national territory, with the aim of creating development areas, is particularly interesting.

The careful investigation we have so far followed, can only lead us to the consideration that the most efficacious demonstration of the importance of changes in economic dynamism is the economic history.

Stagnation or evolution of the economy, in static or developing phases that the various human societies undergo, rejoins itself with those profound changes in the most disparate fields of human activity and in the natural events so far outlined. Primitive history cannot but begin with a description of hunting practices and the gathering of fruits of the earth undertaken by our earliest ancestors. From the primordial exercise of these activities, these groups of men who led a primitively organized, elementary existence, moved onto agriculture and the animal rearing. From tribal communities urban settlements begin to gain a somewhat unsteady footing. Animal taming and the practice of agricultures involve long evolutionary processes that are still taking place today in animals and plants.

This flux of biological changes provoked by man, begun almost instinctively through a rudimentary selection and continues today with the aid of scientific cognitions. When we think of sea cultures, of the exploitation of sea weed as future sources of nourishment, we remain in the furrow ploughed by man at the dawn of civilization. Thus, the earliest history of man is essentially an economic History.

Discoveries must, of course, have played a fundamental role in the opening part of this adventure. Fire, the wheel, the first rudimental agricultural implements, the earliest signs that may be put to use, these are conquests obtained during a vast span of time and represent what great importance was attached to any form of novelty in such primitive context. Without those difficult and painful acquisitions, human civilizations would never have reached the present degree of evolution.

ENTERPRISE AS AN EXPEDIENT

Granted that the term enterprise is a generic one which covers ultimately almost all human activities directed at a specific end, we believe that even the most modest activity (aimed at achieving some benefit for those involved) can be included in the word throughout our work. However, we do not wish to include the smaller agricultural undertakings in this term. Clearly, a direct farmer is an entrepreneur, as is a craftsman who operates alone or with the sole aid of his family. We would rather remind ourselves that there are some very particular individual types of enterprises. Often we see them operate and do not think of them as economic enterprise that have an importance beyond the one represented by the minimal trade they are involved in.

When we see a man in a street opening an umbrella, setting it the ground, placing ties on it and attempting to sell them to passer-byes, we must think that we are confronted with a brave and ingenious entrepreneur. When we see a puppet theatre in a park and two characters moved by the puppeteer and interpreted by the voice of the same man, we are confronted with an entrepreneur which almost certainly has a considerable artistic tradition behind him that represents a much more solid background to that of successful improvised industrialist who merely rides the waves of favourable circumstances. When we watch a man behind the counter of a refreshments kiosk, we must think that he has gone through a lot of problems to obtain a licence, permission to occupy a public place, authorizations and visas of all kinds; in short, that he has had to struggle against bureaucracy. Here too there is a tradition and an ability that not all those involved in similar activities can possess.

The choice of place where one ought to locate one's business requires a developed sense of commercial intuition and very particular equipment. Arrangements with refreshment, syrup and beer suppliers are the product of laborious transactions that test the individual's capacity to argue and successfully exploit competition.

In this sphere of activity, we can see how cash may be skillfully kiosk owners, placed in favourite tourist spots, can take advantage of the differences in currency exchange rates.

What has been said, relates back to an economic law expressed in a variety of ways and under certain determinations: we rather refer to it as the law of the wheel. The small entrepreneur gives rise to an enterprise that develops, acquires complexities, increases its management costs, while still maintaining a superior position in respect of medium and small undertakings. The only valid competitor is the smallest of entrepreneur who minimizes all general expenditures to the full; being extremely flexible, he constitutes the only possible competition, not so much by what he is, but by what he could become.

This modest individual enterprise is the only true seed of novelty and development; it represents one of the most solid guarantees for the replacement of social classes. Not by chance men who have established great enterprises are men of very humble social extraction and these are the men that succeed in the moments during which commitments and difficulties require that gift for improvisation and fast thinking that is lacking in those concerns already set in their ways and enclosed in a rigid context.

The law of the wheel can only direct enterprise back a biological and natural principle. Enterprise is born, develops becomes complex and involuted, grows old and declines. This cycle is similar to that experienced by man: enterprise is made by men and is founded on men.

Enterprise, however, has an individuality of its own, lives its life, grows older and withers away because in the long run it is uncapable of adapting to continuous and ever swifter changes; every reality is eventually superseded, every form carries in its existence the cause and effect of its dissolution.

Chapter nine

THE DECISION

THE PROTAGONIST'S IDENTITY

We have identified in man, the protagonist in business management, the object and the end of enterprise, as we have understood it, projected in the future, beyond political forms that the society may assume, even within the inevitable modification of instruments and methods.

We now deem it necessary to add to the observations already conducted an ulterior verification of the identity of man as an individual.

In a comparison between cybernetic instruments and the human mind we have already pointed out the outline we wish to trace. Man as individual may be considered under different profiles that contribute to form the constellation of innumerable characteristics which creates the irripetible phenomenon of individuality.

The constitution: we now enter a specifically genetic field. By constitution we mean the exterior form that each individual presents of his organism as a whole, in other words that which is aptly defined as the *habitus*. The morphological variations are infinite because the various constitutions never exist on their own; they constantly interact with one another. One may talk in terms of short limbed or tall types, following a classification that has changed nomenclature through the ages, but which dates back to antiquity. The dream to associate temperament with physical presence in an effort to discover in the general lines of the body a correspondence for the inclinations and tendencies of the subject, by us, considered, is just as ancient.

What one really wishes to be able to determine is the temperament. By temperament is meant the whole complex of characteristics in the nervous system, both central and vegetative, in a given individual. According to his nervous system, man will possess different degrees of sensibility and reflexes to external stimuli; so we have hyperexcitable types and extremely calm, relaxed individuals. As we already mentioned, constitution and temperament form a human type and from Hippocrates to Pende typology has articulated itself in the most varied nomenclatures and exceptions. Obviously, the types analysed by this theory have a purely indicative value, while each individual comprises a series of characteristics relative to more types.

Another element useful for an indication is character, term with which we mean to refer to the most elect part of man's psychic faculties, the functions of the spirit. Character is determined by genetic inheritance but on top of being the fruit of the constitution and temperament, it is also determined by life's conditions; in other words by the experiences achieved in his past, and especially in the first years of life. The concept of character, even if throughly understood does not possess absolute characteristics; in it, several elements like will, sensitivity, emotivity, degree of psychic alertness, come into play. Tone is also part of character: a high, vital tone characterizes the optimist and active being, as well as the ambitious and megalomaniac one; a low vital tone is a prerogative of the undecided, the renouncer. Notwhitstanding the fact that opinions may come to blows on this subject, we feel that intelligence must be associated with character, in order not to give an uncomplete picture of the psychic individuality of each person.

Psychology has been defined as the study of behaviour. Given the importance that the behavioural study of man has assumed in business matters, it explains the interest we have manifested in relation to the importance of psychological movements.

We must, nevertheless, qualify this by affirming that these studies are directed more towards a strumentalization of man in the attempt at maximizing his performance, rather than modify, in a human sense, the structures within he operates.

Recently, a new science has begun to observe more closely a particular form of human behaviour; we are referring to kinesics which studies and researches the behavioural models of non verbal communication. It interprets the significance of gestures and attitudes, often unrealized, with which man expresses moods, desires, rebellions that contradict his verbal messages.

Territorial instinct, which has been studied in the past few decades by illustrious scientists, and to which has been credited a prioritarian value with due respect to the sexual instinct, finds unsuspected manifestations in man's behaviour in everyday living, and a particular exemplification in the business realm.

We thus arrive at various conceptions of space: the intimate close-up zone, the personal distance zone, social distance zone, public distance zone. This is not the place to enter into details, however, we may refer to the distance between employees in offices; between a public speaker and the crowd; between director and employee; to the shapes and sizes of the big bosses' desk.

We may arrive at a distantial dynamics during a conversation on work matters through which from a formal phase — during which we remain distanced by a table lenght — we arrive at a closer distance in which negotiations take place on two nearby armchairs. This, evidently, stands for a break-through of initial diffidence, or — at least — that we wish to convey to one another a certain confidence and trust.

Head and leg movements, the playful displacements of objects, are all manifestations of a silent language that only a long familiarity may interpret, but the existence of which cannot escape the careful observer.

We have thus arrived at the motivation; let us recall some concepts that possess an indicative value on what is meant here by motivation, especially in the context of the anglo-saxon management science. Motivation derives from elements such as success, responsability, hierarchical carreer, gratification. The opposite of motivation would then be defined as lack of satisfaction; performance at work enters in relation to both motivating conditions and maintenance requirements. Maintenance requirements are inherent in the necessities (physical, environmental, economic security, particular indemnity-rights derived by seniority).

Meeting maintenance requirements has been judged as the least satisfying task when compared with the result achieved in more gratifying forms of employement. In other words, it is more profitable for business to intervene on the given person by encouraging him with a greater degree of responsability, access to a wider range of informations, with a greater freedom of action, rather than intervene on those aspects which we have included in the termed maintenance requirements.

Since we are talking about management science — and thus we are referring ourselves to individuals giften with the capacity to make decisions — we may mention some motivating examples together with attitudinal characteristics that may be inherent in such men: the quest for success, the thirst for power and other variants of these fundamental notions, that may be discerned from what we have already mentioned.

Whoever possesses a strong drive for success, does not only worry about the goal proposed, but also of the best way to achieve it. A good manager is never a gambler; he prefers circumstances in which the success of given operation depends upon personality and effort and not on chance and other external elements. Yet he reveals a strong fascination for calculated risk. It is as if he desires to have difficulties constantly barring his path; he may complain about them but at the same time they bring out the necessities inherent in his spirit to overcome all obstacles and thereby achieve the ultimate gratification. It is significant that he should wish to overcome unpredictable obstacles which are, nevertheless beyond his capability for action. A man who is spurred on by success has a great need to know wheter the results acquired have been truly positive or not. Clearly, then, he is interested in activities centred on results.

The analysis of power-driven individuals, is more difficult; we may find ourselves confronted with expressions of cold indifference towards others, as we may witness the concretization of a power drive in the wish to heel « humanity ». Often, highly humanitarian and altruistic manifestations are none other than the sublimation of an aggressive spirit; from similar genetic characteristics and psycological schemes wholly different lives may develop.

The need for affiliation — that is the need to be popular and well respected — denotes consideration for other people's feelings and the desire for friendly relationships. The individuals who have these characteristics are more suited to a career in internal and external public relations within business concerns than in activities concerned with decisional commitments.

But the behaviourist science does not apply its discoveries only to normal human behaviour, but also to the more abnormal ones; naturally, in this context, we limit the meaning of abnormality to a small diversion from the already hard-to-define concept of normality. An abnormal behaviour must, obviously, be observed within the context afforded by specific cultures or ethnic groups, as well as those particular moods in which an individual may find himself — anxieties, preoccupations or other moods — that may cause deviations from natural behaviour.

By adopting a technical vocabulary to indicate moderate pathological conditions, that are becoming increasingly more common, due to almost epidemic states of tension and anxiety, we may make useful distinctions between various types of personality. At the two extremities we have extroverted types that consider their activity as having extraordinarily importance, and the remissive types that tend to underrate and inhibit themselves. The less an individual is dominated by self esteem, the less he is intimidated and less he attemps to fulfill other people's expectations the more he will be able to allow his inborn capacities to blossom.

SCIENTIFIC THOUGHT AND MANAGERIAL PRAXIS

Medieval thought, in all its aspects, was based to preceding eras, even if influences from far away countries like India and China must not be forgotten. Certain ideas proclaimed by French Illuminists are deeply rooted in Chinese thought. But Scholasticism founded all formulations on a logic that refers back to the Bible, Aristotle and the Fathers of the Church. The figure and work of Thomas Aquinas are highly emblematic of a spiritual attitude rather than simply an era.

The great changes owed to technical progress, the discovery of new lands, gave rise to a new way of looking at science, that gradually adopted a system empirical data. Naturally, the old theoretical attitude, in both the idealist philosophy and catholic tomism survives the however strong, new, empirical, and pragmatic approach.

But empirical thinking has developed within scientific research until it influenced determinantly the French Encyclopedists.

Along with the encyclopedists we may remember Comte — to whom we owe the birth of sociology — and Spencer's incalculable contribution.

These authors based themselves on a logical organization of essentially empirical notions. The empirical method had already been adopted by Leonardo da Vinci as a working proposition for the solution of an enormous number of technical and scientific problems.

This short diversion finds justification in the relevance it will have on business management. In fact, the scientific approach manifests itself in every sphere of human existence and thus it will also be pertinent to an industrial, management science.

A divergence of attitudes, then, verifies itself in a wide number of analogous situations to the dichotomy we encountered earlier between a theoretic and a pragmatic approach. There is nothing extraordinary in this: it concerns a new science and it is therefore inevitable that those interested in it should choose, between the two, the more congenial one according to one's inclination and personal cultural training.

Secondly, there exist such a great wealth of conflicting works and theories on which programming and managerial choices should be based. It is clear, then, that it is difficult to place the foundations for decisions in experience, because it requires long years of practice, on top of the obvious obstacle constituted by the swift changing situations and the whole context in which we operate.

We believe that in the business field it is necessary to be allowed complete freedom of judgement, in the sense that theories must be a support to decisions rather than a hinderance.

Even if the discipline of managerial thought is still at an embryonic stage it exercises un unquestionable charm and it is already an authoritative presence in the practical management of business concerns and great organizations. This presence, however, must have a stimulating value and not become a source for dogmatic attitudes. In business praxis one needs to conduct research, examination, discrimination, on the basis of concrete facts after all possible data have been collected from the actual and fascinating internal and external business reality.

We have noticed how vitally important it is to formulate the problems in exact terms, and the questions we consequently deem necessary to ask in a precise manner; all too hasty or approximate elaboration of facts would, in fact, lead to answers that call for all too imprecise decisions that fulfill a small number of clauses in the problem's actual conditions.

On the social plane we must assume an attitude that extends the empirical methodology employed in scientific research, even to the humblest realities of everyday life. In order to do so, we must examine everyday activities, each individual's work, with due respect and consideration in the awareness that cultural and scientific research possess a dimension in which each of the collectivity may find some comfort.

It is essential that everyone apprehends scientific method as a mental form which allows man to discover and employ the simplest and most convenient systems of work. If this concept is, as it should be, widely accepted one would find a general slendering of all useless procedures that infiltrate the social tissue, hindering and disturbing the free flux of all human activities.

It is pointless for science to astonish and astound us with space conquests, with spectacular medical transplants, when the individual is forever harassed by the necessity to resolve the small and artificial problems produced and perpetrated by society itself. If men, as individual beings will become more alert and intelligent in the way of thinking and acting, society will, by necessity, become a gradually less hostile entity for the individual. A more relaxed environment would surround us, devoid of those frictions that are often caused by misunderstandings of one another's problems rather than by antithetical viewpoints.

Returning to our most specific concern, we notice that a great deal of factors are to be held responsible for the different methodological approaches followed in business research in the United State and our country. We shall attempt to name but a few.

They are in the first place « cultural factors » (cf. Stafford Beer). Management culture, as such, derives from religious, historical, philosophical components. Among these, the most consequential appears to be the religious one. It is not worth listing some of the effects produced by it on the economic production and hence business texture, as it would prove a far too repetitive process.

Among the cultural factors, the most relevant ones are, undoubtedly, faith in budgetings, efficiency, the scientific management of business. They have led to an always greater differentiation between the two schools of thought: the italian one, appears to us, to avail itself — in an appropriate measure — of the deductive and inductive methods; the american one, on the other hand, by abandoning the characteristic pragmatism of its culture, has often chosen the path of abstract induction, of efficiency. The latter, above all, has been made the object of eccessive worship by american businessmen like P. F. Drucker who holds that efficiency is a habit, that is, a praxis which can always be taught and, hence learnt. For Drucker, this praxis must be mastered in the same way as we are made to master, by memorizing ad nauseam, mathematical tables; in other words it must become an automatic and conditioned reflex, perpetrated by constant practice.

This form of pure and abstract efficiency may make one disregard the concrete problems a managing director has to face: hence the need for an operative research to help all managers, by providing models that allow them to take decisions on an all too sophisticated statistico-probabilistic basis. This cannot suffice: efficiency in production is also necessary in a business; hence the birth of scientific organization of work, of taylorism, of industrial relations, studies.

Secondly, we are dealing with « juridical factors ». The United States are endowed with a capillary and extremely effective system for tax collection. Taxes on enterprises' profits are based on an estimated budget certified by special organs; said certified budget it also employed for the determination of profit distribution. All this ensures that the financial balance is employed as basis for an analysis of the enterprise's condition from an economic and financial viewpoint. Naturally, a technique for balance analysis came into being. Said method is widely employed in the United States but cannot be introduced in our country because of the laughable value of financial balance declared by Italian concerns. We may, thus, catch a glimpse of which and how large the problems that have led to two such distant positions among scholars of an identical object of research — that is business — are.

Yet, perhaps, the most important factor in this diversity of methodological approach is the *dynamic factor,* meaning the sum total of events that have followed one another in a time lapse of a *generation* of managers and that have contributed to the latter's professional, and psycological formation. One such element is war. Europe has suffered innumerable wars; our nations have always had to live through, and be submitted to, this traumatic experience, and maybe this explains, perhaps as a biological inheritance, the total and widespread disillusionment with what may be in store for us in the future. The opposite standpoint is commonly held in the United States, where the future is regarded as an explained, or at least explainable, variable that can be introduced in any business pattern.

This is the greatest obstacle to the introduction of techniques akin to a scientific management of business concerns; operative budget, rotative budget, forecasts of cash flow for the analysis of investments; all is based on estimates, on predictions for the future, and possesses the overall flavour of an artificial concoction based on nothing, or worse, on an illusion. These procedures are followed by our managers too, but without the ostentatious faith that drives our american colleagues relentlessly on.

CONCLUSION

Finally, we would like to sum up the observations we have carried out with regard to business management.

At the basis of our beliefs lies the reality of an economy in constant mutation. While talking about changes, we wish to emphasize that we are not conjecturing the wake of stagnation periods; we merely wish to indicate the reality of an epoch that is characterized by the constant changes that occur in each and every field. Everyone shares this concept and all are agreed that public and private structures find themselves in truly « emergency situations ». Very often, however, the theoreticians, for motives inherent in their function, hypothesize a static reality, or at least they examine it statically, in order to be able to concretize indexes, models and research patterns. We understand the importance of theoretic research and the necessity that it constitutes the cultural substrata of those who are, or will be, called upon to manage business concerns and map out plans and programmes. On the other hand we also believe in the most thorough and complete examination of a concrete case, enriched by the verification of numerous hypotheses that may have been worked out.

Working hypotheses must be suggested by the knowledge of practical problems; only when these problems are lived through in a direct participation and with specific responsibilities, one may be aware of the availability and suitability of possible alternatives. Those who really feel the burning necessity to resolve (and soon) a certain problem because it affects the efficiency and possibly the very life of the business concern, may usefully employ the specific cultural experience, without loosing himself in useless conjectures and improbable hypotheses. Clearly, then, the man or the men who have to take decisions must acquire a mental state that is influenced by a knowledge in depth of managerial science but also by a verification achieved through constant practical experience in everyday business activity.

In spite of widespread specialization, in spite of the dimensions of the group within which a man is placed, it is important, in the business world as much as anywhere else, that he maintains and safeguards his individuality since man is the only element capable of working out solutions that are efficient as well as congenial to the possibilities and nature of human beings.

If, the decision is arrived at through a collaboration or a compromise between various individuals it can only reinforce, instead of weakening, our thesis. We are calling for men as individuals to run all enterprises and not for demiurges be they men or plain elaborators. The times of the great charismatic figures, the enterprising generals, have disappeared into oblivion, in spite of the fact that certain rituals are mantained and perhaps exhasperated in the bureaucratic system where inefficiency reigns unperturbed.

The cult and reverence for form and ceremonials to which man has always kneeled before, is not yet extinct; to support our assertions we may remind the reader of two very different but equally interesting works: Graham Cleverley's «Managers and Magic » with its paradoxical and ironic manner, though accurate inside business complexes, and Marc Bioch's « Les rois thaumaturges » a book with very different aims but characterized by a wholly special approach geared to demonstrate just how deeply man has been enticed by the lust for power.

Yet, whether we are dealing with an individual or a group of individuals, according to the dimensions of the context in which they operate, it is clear that the aids of an instrumental, mathematical and sistematic nature will prove useful only after the specific problem has been wholly organized in its practical and material aspects. Direct observation is also elaborated at a subconscious level, yet this rarely happens when confronted by a schematization of reality.

The real elements and men that form the object of the

problem communicate with his sensitivity as well as his brain aiding him in that quest toward concrete truths that are inextricably inherent in operative decisions. Knowledge and practical observation are thus the required conditions that will allow one to arrive at these concrete truths and the flexibility of spirit necessary for managerial decisions.

We therefore need to conjecture what the collection of available objective data and the estimate result of the influences inherent in given decisions, may lead toward in the future. The aim must consequently be to represent the most likely future we shall have to be dealing with.

We have also been aware of a wider context, but the characteristic element of our research is, above all, a business one; we therefore wish to work our way to the end by evaluating the future prospects in said field. Not fortuitously, we wish to exchange the term business concern for enterprise.

If, in fact, a business concern is the complex of organized goods for the fulfillment of enterprise, enterprise is, on the other hand, a form of activity: the relationship between the two is similar to that of means to ends.

Enterprise is thus an end, and as such, we wish to contrast it with the patrimonial concept of a business concern. Enterprise, in the widest connotations of the word cannot but include all human activity.

Even if we limit ourselves to a stricter vision of the term, as an organized activity for the production of goods and services, it still remains a vast field of human interest.

Enterprise has for its goal the satisfaction of man's needs, but it cannot be restrained by this. In order for it to be made in man's measure, it must not only fulfill material necessities with its products and services, it must also communicate to man the flavour of its usefulness and hence its dignity. Without a job that is not exempt, though not wholly made up of, intellectual and physical effort, man cannot feel a dignified member of the collectivity.

In order to satisfy these requirements, enterprise must not assume gigantic proportions. If we choose to adopt a purely economic criteria, great dimensions may constitute a considerable advantage, but if we wish to carry out a primary examination of the so called social indicators, we must think of future enterprise as being characterized by medium size entities.

It is impossible to establish absolute criteria, given the wide diversification of the various activities, but it is necessary for all those involved in enterprise to have a sufficiently approximate idea of its true aim, of their own as well as their colleagues' roles, of all the aspects of its existence. The only possible motivation is the awareness of what one may contribute to its progress and the problems that derive from the actualizations of the proposed ends. In a huge enterprise, this is rarely achieved since it relies too much on greatly specialized and detached units.

Logically, when one talks of enterprise, in our context, one does not refer to the juridical aspect that it may assume, quite the opposite, one wishes to emphasize the fact that enterprise is a condition of human activity which is destined to survive through any possible socio-economic context. We believe and hope that the characteristics of enterprise in the future will come into being especially through the possibility of giving increasing importance to the relationships between internal factors since the condition for the realization of a complete life lie within it. Enterprise must thus be integrated by extra-economic factors capable of redimensioning the importance of individuals working within it. This is a kind of redimensioning that expresses itself with greater importance and consideration for human work and for the worker as such.

In order for it to happen, certain premises of a political and socio-economic character, without which we would witness a regression toward a hidden paternalism or a simple research on motivations according to the philosophy of modern management, must be fulfilled. Yet, it is not only necessary to resolve the political situation, but also that the workers-consumers modify their attitude toward the social reality by beginning to distinguish between economic and other necessities.

Each individual must understand how indispensable it is to give up indiscriminate consumism of economic goods as the method leading to instant satisfaction and self-achievement. If each single individual does not practice looking at oneself and others according to value judgements outside the economic scale he will only remain enslaved by a system that will be perpetrated by his very choices.

A consumer society that places consumption as only means for satisfaction and qualification, is a society that will never be able to reach greater levels of freedom and dignity than the present ones. It will, in fact, have chosen not liberation from need but consumism as its principal objective and consequently it will pay the ever increasing price in terms of growing enslavement to the present system.

If it is true that men have the government they deserve, we may affirm that the choice for a qualitative or quantitative standard of living in the future will be the determinant factor for the kind of society in which man will decide to exist.

BIBLIOGRAPHY

AGNESSE P., *Clef pour l'écologie,* Seghers, 1971.

ANTHONY R.N., *Planning and Control Systems,* Harvard University, 1965.

ARDREY R., *African Genesis - A personal investigation into animal Origines of the Nature and Man,* Collins, 1961.

BARBER J.W., *A History of economic Thought,* Penguin, 1967.

BASSETTI P., *Strategia regionale per l'uomo e per l'ambiente* (from « *Quaderni di San Giorgio* » n. 34), Sansoni, 1973.

BEER S., *Brain of the Firm,* Penguin, 1972.

BENTIVEGNA V., *Il problema dell'inquinamento nell'economia* (from « *La città inquinata* » *Saggi di Scienza e Tecnica*), Medicea, 1972.

BETTELHEIM G., *L'économie allemande sous le nazisme,* F. Maspero, 1971.

BIOCH M., *Les rois thaumaturges,* Leclerc, 1961.

BONNEFOUS E., *L'homme ou la nature?,* Hachette, 1970.

BRUNETTI CODA, *Introduzione alle analisi di bilancio,* Libreria Universitaria Editrice, 1972.

BÜHLER CH., *Psycologie im Leben unserer Zeit,* Droemersche, 1962.

CAO PINNA V., *Le prospettive dei consumi alimentari in Italia,* Giuffrè, 1962.

CARLSON P.G., *Quantitative Methods for Managers,* Harper & Row, 1967.

CETRANGOLO E., *Introduzione a Lucrezio « Della Natura »,* Sansoni, 1969.

CH'U CHAI and W. CHAI, *The Story of chinese Philosophy,* Washington Square Press, 1961.

CLEVERLY G., *Managers and Magic,* Longman, 1971.

COMMONER B., *The closing Circle,* Knopf, 1971.

CONTI C., *Teoria delle misurazioni per le scienze sociali,* Mazzotta, 1972.

COTTINO G., *Entry S.p.A.* in « *Novissimo Digesto Italiano* ».

CURRY K., LINDAHL, *Conservation of Survival - An ecological Strategy,* Lindahl, 1973.

D'ALBERGO E., *Economia della finanza pubblica,* Giuffrè, 1971.

DAJOZ R., *Précis d'écologie,* Dunod, 1971.

DA RIVA D. - RICCI G., *Elementi di organizzazione delle informazioni,* Franco Angeli, 1972.

DEANE PH., *The first industrial Revolution,* Cambridge University Press, 1967.

DE FINETTI B., *Un matematico e l'economia,* Franco Angeli, 1969.

DE MARIA G., *Trattato di logica economica,* Cedam, 1966.

DE MARIA G., *I grandi sistemi coercitivi,* Cedam, 1969.

DERNBURG T.E., MC DOUGALL, *Macro-economics: the Measurement Analysis and Control of Aggregate economic Activity,* Mc Graw Hill, 1963.

DOBB M., *Wages,* Nisbet, 1956.

DOCKES P., *L'espace dans la pensée économique du XVIème au XVIIIème siècle,* Flammarion, 1969.

DRUCKER P. F., *Managing for Results,* Heinemann, 1964.

DRUCKER P. F., *Russian economic Development since the Revolution,* Routledge, 1959.

DRUCKER P. F., *The effective Executive,* Heinemann, 1967.

DRUCKER P. F., *The Practice of Management,* Heinemann, 1955.

DYNAMICS SYSTEM GROUP (M.I.T.), *Toward a global equilibrium - The Limits of Growth,* Allen Press, 1973.

EINAUDI L., *Prediche inutili,* Einaudi, 1974.

FACCHINI PANCERA, *Dipendenza economica e sviluppo economico in Israele,* Franco Angeli, 1973.

FANFANI A., *Storia economica (vols. I-II),* Utet, 1968-1972.

FAURE R., et al., *Mathématiques générales,* Dunod, 1968.

FAURE R., et al., *Mathématiques nouvelles,* Dunod, 1970.

FERRAROTTI F., *La sociologia,* Garzanti, 1967.

FERRERO G., *Le analisi di bilancio,* Giuffrè, 1972.

Forte C., *Ecologia ed economia urbana* (from. « *La città inquinata* » *Saggi di Scienza e Tecnica*), Medicea, 1972.

Foster C.D., *The Transport Problem,* Blackie & Son, 1963.

Galbraith J.K., *Economics and the public Purpose,* Deutsch, 1974.

Galbraith J.K., *The affluent Society,* Hamilton, 1958.

Gandolfo G., *Metodi di dinamica economica,* Isedi, 1973.

Georgescu N. Roegen, *Analytical Economics,* Harvard University Press, 1966.

Georgescu N. Roegen, *Economies et Sociétés,* Droz, 1970.

Georgescu N. Roegen, *El Trimestre Economico,* XXXIV, 1967.

Georgescu N. Roegen, *Distinguished Lecture Series N. 1, The Entropy Law and the economic Problem,* The University of Alabama, 1971.

Gerelli E., *Economia e tutela dell'ambiente,* Il Mulino, 1974.

Gerschenkron A., *Economic Backwardness in historical Perspective,* Harvard University Press, 1962.

Giani C., *La regolazione automatica nell'industria,* Zanichelli, 1965.

Gilbert B., *Il linguaggio degli animali,* Mondadori, 1972.

Gramsci A., *Note sul Machiavelli,* Editori Riuniti, 1972.

Harlow, Mc Gaugh, Thompson, *Psychology,* Albion, 1971.

Heichelheim F.M., *An ancient economic History,* Sijthoff, 1970.

Heller R., *Management in Britain: the real Challenge,* Ashridge, 1967.

Hicks J.R., *Capital and Growth,* Oxford University Press, 1965.

Hicks J.R., *Capital and Time: an Austrian Theory,* Clarendon, 1973.

Hyams E., *Killing no murder,* Panther, 1970.

Hyams E., *Plants in the service of Man - Animals in the service of Man,* Panther, 1972.

Istituto Gramsci, *Uomo, natura, società, ecologia e rapporti sociali,* Editori Riuniti, 1972.

Klein R.L., *The Keynesian Revolution,* Mac Millan, 1952.

Kloter P., *Marketing Management: Analysis, Planning and Control,* Prentice Hall, 1967.

La Francesca S., *La politica economica del fascismo,* Laterza, 1973.

La Malfa U., *La Caporetto economica,* Rizzoli, 1974.

Lesourd G., *Histoire économique: XIXème et XXème siècles,* Colin, 1963.

Levinson Ch., *Capital, Inflation and the Multinationals,* Allen & Unwin, 1971.

Lewis T., Taylor L.R., *Introduction to experimental Ecology,* Academic Press, 1967.

Likert R., *The human Organization: its Management and Value,* Mc Graw Hill, 1967.

Lipsey R.G., *An Introduction to positive Economics,* Weidenfield & Nicolson, 1971.

List F., *Il sistema nazionale di economia politica,* Isedi, 1972.

Lombardini S., *Monopolio e concorrenza nella politica economica,* Franco Angeli, 1973.

Luzzato G., *Storia dell'età moderna e contemporanea, vols. I-II,* Cedam, 1955-1960.

Maldonado T., *La speranza progettuale,* Einaudi, 1971.

Marx K., *Il Capitale, Libro III,* Editori Riuniti, 1972.

Massa G.B., *Il controllo di gestione,* Etas Kompass, 1971.

Mattick P., *Marx and Keynes, The Limits of the mixed Economy,* Porter Sargent, 1969.

Morice E., *Dictionnaire de statistique,* Dunod, 1967.

Morison E., *Men, Machines and modern Times,* M.I.T., 1966.

Morris D., *The naked Ape,* Jonathan Cape, 1967.

Munford L., *In the Name of Sanity,* Harcourt Brace, 1954.

Myer J.N., *Financial Statement Analysis,* Prentice Hall, 1961.

Nadler G., *Work System Design: the ideal Concepts,* Richard Irwin, 1967.

Napolitano G., *Il consumismo,* Almanacco, 1971.

Nebbia G., *Ecologia ed economia,* Cedam, 1973.

Oppenheim F., *L'Ecole du profit,* Ed. du Seuil, 1970.

Packard V., *Hidden Persuadors,* Longman, 1957.

Packard V., *The Status Seekers,* Mc Kay, 1959.

Paillard, Bloch, Piecon, *Traité de psychologie expérimentale,* Presses Universitaires de France, 1966.

Perroux F., *L'économie du XXème siècle,* Presses Universitaires de France, 1964.

Perroux F., *Indépendence de l'économie nationale*, Aubier Montaigne, 1969.

Prince R. Th., *Information System for management Planning and Control,* Irwin, 1966.

Procacci G., *Introduzione a Machiavelli: Le Opere,* Editori Riuniti, 1969.

Pullani E., *Lo sviluppo multinazionale delle imprese industriali,* Etas Kompass, 1973.

Revel J.F., *Contro censures,* J.J. Panvert, 1966.

Revel J.F., *Ni Marx ni Jésus*, Robert Laffont, 1970.

Ronchey A., *La crisi americana,* Garzanti, 1975.

Ruffolo G., *Rapporto sulla programmazione,* Laterza, 1973.

Russell B., *The ABC of Relativity,* Allen & Unwin, 1958.

Sabattini M., *Fortran Cobol PL/1: Guida alla programmazione dell'elaboratore elettronico,* Etas Kompass, 1969.

Saraceno P., *La produzione industriale,* Libreria Universitaria Editrice, 1973.

Sauvy A., *Croissance zéro?*, Calmann-Lévy, 1973.

Seger I., *Knaurs Buch der Modernen Soziologie,* Droemersche, 1970.

Servan-Schreiber J.J., *Le défi américain,* Denoël, 1967.

Siskind C.S., *Electrical Control Systems in Industry,* Mc Graw Hill, 1963.

Società italiana economisti, *Prezzi e produttività,* Utet, 1967.

Tagliacarne G., *Tecnica e pratica delle ricerche di mercato,* Giuffrè, 1960.

Tagliacarne G., *La carta commerciale d'Italia,* Giuffrè, 1968.

Toffler A., *Future Shock,* Bodley Tead, 1970.

Torresan C., *Automazione di impianti chimici e termici,* Hoepli, 1974.

Trabucchi A., *Istituzioni di diritto civile,* Cedam, 1971.

Turner J.S., *The chemical Feast,* Ralph Nader Group, 1970.

Vacca R., *Il medioevo prossimo venturo - La degradazione dei grandi sistemi,* Mondadori, 1972.

Volpato M., *Studi e modelli di ricerca operativa*, Utet, 1971.

von Bertalanffy L., *General System Theory,* George Braziller, 1968.

Wolf R.P., *In Defence of Anarchism,* Harper & Row, 1970.

Yamane T., *Mathematics for Economists,* Prentice-Hall, 1968.

Zappa G., *Il reddito d'impresa,* Giuffrè, 1950.

Printed in Italy in may 1975
by Arti Grafiche Milillo, Roma